T0269157

CAMBRIDGE LIBRARY COLLECTION

Books of enduring scholarly value

Earth Sciences

In the nineteenth century, geology emerged as a distinct academic discipline. It pointed the way towards the theory of evolution, as scientists including Gideon Mantell, Adam Sedgwick, Charles Lyell and Roderick Murchison began to use the evidence of minerals, rock formations and fossils to demonstrate that the earth was older by millions of years than the conventional, Bible-based wisdom had supposed. They argued convincingly that the climate, flora and fauna of the distant past could be deduced from geological evidence. Volcanic activity, the formation of mountains, and the action of glaciers and rivers, tides and ocean currents also became better understood. This series includes landmark publications by pioneers of the modern earth sciences, who advanced the scientific understanding of our planet and the processes by which it is constantly re-shaped.

Elements of Geology

Between 1830 and 1833, Charles Lyell (1797–1875) published his three-volume *Principles of Geology*, which has also been reissued in this series. The work's renown stems partly from the fact that the young Charles Darwin, on his voyage around the world aboard the Beagle, became influenced by Lyell's ideas relating to gradual change across large spans of time. Shaping the development of scientific enquiry in Britain and beyond, Lyell was determined to disconnect geology from religion. He originally intended some of the present work, first published in 1838, to be a supplement to the *Principles*, but later expanded it to serve as a general introduction to geology. The topics covered include the formation of various rock types, matters of field geology, and how the presence of marine fossils above sea level could be explained by the land rising, rather than the sea retreating. Many salient points are illustrated with woodcuts.

Elements of Geology

CHARLES LYELL

CAMBRIDGE
UNIVERSITY PRESS

CAMBRIDGE
UNIVERSITY PRESS

University Printing House, Cambridge, CB2 8BS, United Kingdom

Cambridge University Press is part of the University of Cambridge.
It furthers the University's mission by disseminating knowledge in the pursuit of
education, learning and research at the highest international levels of excellence.

www.cambridge.org
Information on this title: www.cambridge.org/9781108075091

ELEMENTS

OF

G E O L O G Y.

BY

CHARLES LYELL, ESQ. F.R.S.

VICE-PRESIDENT OF THE GEOLOGICAL SOCIETY OF LONDON, ETC.
AUTHOR OF "PRINCIPLES OF GEOLOGY."

" It is a philosophy which never rests—its law is progress: a point
which yesterday was invisible is its goal to-day, and will be its starting
post to-morrow."—EDIN. REV., No. 132. p. 83. July 1837.

LONDON:
JOHN MURRAY, ALBEMARLE STREET.
MDCCCXXXVIII.

LONDON:
Printed by A. SPOTTISWOODE,
New-Street-Square.

TO

WILLIAM HENRY FITTON,

M.D. F.R.S.

VICE-PRESIDENT OF THE GEOLOGICAL SOCIETY,

ETC.

MY DEAR DR. FITTON,

I HAVE great pleasure in dedicating this volume to you, as a memorial of an uninterrupted friendship of nearly twenty years, during which we have been engaged in the same scientific pursuits. I am also glad to have this opportunity of acknowledging the benefit which my writings have so often derived from your friendly criticism.

I am,

My dear Dr. Fitton,

Yours, very sincerely,

CHARLES LYELL.

LONDON, *July* 26. 1838.

A 2

PREFACE.

Part of the present Treatise was written originally in the form of a supplement to my former work, entitled " Principles of Geology," and was intended for the use of those students who found certain chapters in the Principles obscure and difficult, for want of preliminary information. I afterwards considered that it would not be incompatible with this object to enlarge the Elements into a separate and independent treatise, to serve as an introduction to Geology proper. As I have thus been led on to become the author of two general works on the same science, it may be useful to explain to the reader that these two publications do, in fact, occupy very distinct ground.

In the Principles a systematic account is given of the operations of inorganic causes, such as rivers, springs, tides, currents, volcanos, and earthquakes; the effects of all being particularly considered, with a view to illustrate geological phenomena. The changes also which the organic world has undergone in modern times, the geographical distribution of different species of animals and plants, the causes of their multiplication and extinction, and their first introduction, are discussed, and the various ways in which their remains become fossil in new deposits. The student, who is familiar with this the larger portion of the Principles (comprising no less than five-sixths of the whole), will, it is hoped, more easily comprehend the explanations of geological appearances proposed in the Elements. On the other hand, those who begin with the Elements, the scope of which may be understood by a glance at the annexed table of contents, will follow more easily the meaning of that part of the Principles in which an attempt is made to point out the bearing on geology of the modern changes of the earth, and to which is prefixed a history of the opinions which have been entertained in this science, from the times of the earliest writers to the present day.

The volume, therefore, now offered to the public, is neither an epitome of the Principles, nor an abridgement of any part of that work. In some places, where I thought it desirable to incorporate in the Elements certain passages of the former work, I have not abridged what was previously written, but have expanded it, giving fuller explanations, and additional wood-cuts, in the hope of rendering it more intelligible to the beginner.

Through the kindness of two of my friends I

have been enabled to refer frequently to two works, not yet before the public, Mr. Darwin's Journal of Travels in South America, 1832 to 1836, &c., and Mr. Murchison's Silurian System; the last of which was presented to me complete, with the exception of the maps and plates, and will shortly be published.

Mr. Darwin's Journal was finished, and ready for publication, some time before the printing of my MS. had begun, but is still detained, to the great regret of the scientific world, because it is to form part of a larger work, including an account of the Surveys of Captains King and FitzRoy, in South America.

N. B. The greater part of the woodcuts in this volume, especially those most difficult of execution, are the work of Mr. James Lee, 97, Princes Square, Kennington. The original drawings in Natural History were done by Mr. Geo. Sowerby, jun., 14, Tibberton Square, Lower Road, Islington.

CONTENTS.

PART I.

CHAPTER I.

ON THE FOUR GREAT CLASSES OF ROCKS — THE AQUEOUS,
VOLCANIC, PLUTONIC, AND METAMORPHIC.

CHAPTER II.

AQUEOUS ROCKS — THEIR COMPOSITION AND FORMS OF
STRATIFICATION.

CHAPTER III.

CHAPTER IV.

CHAPTER V.

CHAPTER VI.

DENUDATION, AND THE PRODUCTION OF ALLUVIUM.

CHAPTER VII.

VOLCANIC ROCKS.

CHAPTER VIII.

VOLCANIC ROCKS — *continued.*

CHAPTER IX.

PLUTONIC ROCKS — GRANITE.

CHAPTER X.

METAMORPHIC ROCKS.

CHAPTER XI.

METAMORPHIC ROCKS — *continued.*

PART II.

CHAPTER XII.

ON THE DIFFERENT AGES OF THE FOUR GREAT CLASSES OF ROCKS.

CHAPTER XIII.

ON THE DIFFERENT AGES OF THE AQUEOUS ROCKS.

CHAPTER XIV.

RECENT AND TERTIARY FORMATIONS.

CHAPTER XV.

CRETACEOUS GROUP.

CHAPTER XVI.

WEALDEN GROUP.

CHAPTER XXIII.

ON THE DIFFERENT AGES OF THE VOLCANIC ROCKS.

CHAPTER XXIV.

ON THE DIFFERENT AGES OF THE PLUTONIC ROCKS.

CHAPTER XXV.

ON THE DIFFERENT AGES OF THE METAMORPHIC ROCKS.

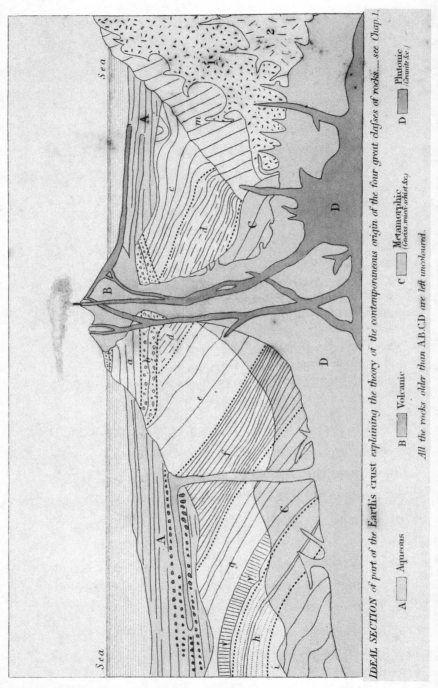

IDEAL SECTION *of part of the Earth's crust explaining the theory of the contemporaneous origin of the four great classes of rocks.—see Chap.1.*

All the rocks older than A.B.C.D are left uncoloured.

| A | Aqueous | B | Volcanic | C | Metamorphic (*gneiss, mica-schist &c.*) | D | Plutonic (*Granite &c.*) |

ELEMENTS OF GEOLOGY.

PART I.

CHAPTER I.

ON THE FOUR GREAT CLASSES OF ROCKS—THE AQUEOUS,
VOLCANIC, PLUTONIC, AND METAMORPHIC.

Geology defined — Successive formation of the earth's
crust — Classification of rocks according to their origin
and age — Aqueous rocks. — Their stratification and imbedded fossils — Volcanic rocks, with and without cones
and craters — Plutonic rocks, and their relation to the
volcanic — Metamorphic rocks, and their probable origin
— The term primitive, why erroneously applied to the
crystalline formations — Division of the work into two
parts ; the first descriptive of rocks without reference to
their age, the second treating of their chronology.

OF what materials is the earth composed, and
in what manner are these materials arranged?
These are the inquiries with which Geology is
occupied, a science which derives its name from
the Greek γῆ, *ge*, the earth, and λογος, *logos*, a
discourse. Such investigations appear, at first
sight, to relate exclusively to the mineral kingdom, and to the various rocks, soils, and metals,
which occur upon the surface of the earth, or at
various depths beneath it. But, in pursuing these
researches, we soon find ourselves led on to con-

B

sider the successive changes which have taken
place in the former state of the earth's surface
and interior, and the causes which have given rise
to these changes ; and, what is still more singular
and unexpected, we soon become engaged in researches into the history of the animate creation,
or of the various tribes of animals and plants
which have, at different periods of the past, inhabited the globe.

All are aware that the solid parts of the earth
consist of distinct substances, such as clay, chalk,
sand, limestone, coal, slate, granite, and the like ;
but previously to observation it is commonly
imagined that all these had remained from the
first in the state in which we now see them, —
that they were created in their present form, and
in their present position. Geologists have come
to a different conclusion. · They have discovered
proofs that the external parts of the earth were
not all produced in the beginning of things, in
the state in which we now behold them, nor in
an instant of time. On the contrary, they have
acquired their actual configuration and condition
gradually, under a great variety of circumstances,
and at successive periods, during each of which
distinct races of living beings have flourished
on the land and in the waters, the remains of
these creatures still lying buried in the crust of
the earth.

By the " earth's crust," is meant that small portion of the exterior of our planet which is accessible to human observation. It comprises not merely all of which the structure is laid open in mountain precipices, or in cliffs overhanging a river or the sea, or whatever the miner may reveal in artificial excavations; but the whole of that outer covering of the planet on which we are enabled to reason by observations made at or near the surface. These reasonings may extend to a depth of several miles, perhaps ten miles; but even then it may be said, that such a thickness is no more than $\frac{1}{400}$th part of the distance from the surface to the centre. The remark is just; but although the dimensions of such a crust are, in truth, insignificant when compared to the entire globe, yet they are vast and of magnificent extent in relation to man, and to the organic beings which people our globe. Referring to this standard of magnitude, the geologist may admire the ample limits of his domain, and admit, at the same time, that not only the exterior of the planet, but the entire earth, is but an atom in the midst of the countless worlds surveyed by the astronomer.

Now the materials of this crust are not thrown together confusedly, but distinct mineral masses, called rocks, are found to occupy definite spaces, and to exhibit a certain order of arrangement.

The term *rock* is applied indifferently by geologists to all these substances, whether they be soft or stony, for clay and sand are included in the term, and some have even brought peat under this denomination. Our older writers endeavoured to avoid offering such violence to our language, by speaking of the component materials of the earth as consisting of rocks and *soils*. But there is often so insensible a passage from a soft and incoherent state to that of stone, that geologists of all countries have found it indispensable to have one technical term to include both, and in this sense we find *roche* applied in French, *rocca* in Italian, and *felsart* in German. The beginner, however, must constantly bear in mind, that the term rock by no means implies that a mineral mass is in an indurated or stony condition.

In order to classify the various rocks which compose the earth's crust, it is found most convenient to refer, in the first place, to their origin, and in the second to their age. I shall therefore begin by endeavouring briefly to explain to the student how all rocks may be divided into four great classes by reference to their different origin, or, in other words, by reference to the different circumstances and causes by which they have been produced.

The first two divisions, which will at once be understood as natural, are the aqueous and volcanic, or the products of watery and those of igneous action.

Aqueous rocks. — The aqueous rocks, sometimes called the sedimentary, or fossiliferous, cover a larger part of the earth's surface than any others. These rocks are *stratified,* or divided into distinct layers, or strata. The term *stratum* means simply a bed, or any thing spread out or *strewed* over a given surface; and we infer that these strata have been generally spread out by the action of water, from what we daily see taking place near the mouths of rivers, or on the land during temporary inundations. For, whenever a running stream, charged with mud or sand, has its velocity checked, as when it enters a lake or sea, or overflows a plain, the sediment, previously held in suspension by the motion of the water, sinks, by its own gravity, to the bottom. In this manner layers of mud and sand are thrown down one upon another.

If we drain a lake which has been fed by a small stream, we frequently find at the bottom a series of deposits, disposed with considerable regularity, one above the other; the uppermost, perhaps, may be a stratum of peat, next below a more dense and solid variety of the same material; still lower a bed of laminated shell-marl, alternating with peat or sand, and then other beds of marl, divided by layers of clay. Now if a second pit be sunk through the same continuous lacustrine *formation*, at some distance from the first, we commonly meet with nearly the same series of beds, yet with slight variations; some, for example, of the layers of sand, clay, or marl, may be wanting, one or more of them having thinned out and given place to others, or sometimes one of the masses first examined is observed to increase in thickness to the exclusion of other beds.

The term "*formation*," which I have used in the above explanation, expresses in geology any assemblage of rocks which have some character in common, whether of origin, age, or composition. Thus we speak of stratified and unstratified, freshwater and marine, aqueous and volcanic, ancient and modern, metalliferous and non-metalliferous formations.

In the estuaries of large rivers, such as the Ganges and the Mississippi, we may observe, at low water, phenomena analogous to those of the drained lakes above mentioned, but on a grander scale, and extending over areas several hundred miles in length and breadth. When the periodical inundations subside, the river hollows out a channel to the depth of many yards through horizontal beds of clay and sand, the ends of which are seen exposed in perpendicular cliffs. These beds vary in colour, and are occasionally characterized by containing drift-wood or shells. The shells may belong to species peculiar to the river, but are sometimes those of marine testacea, washed into the mouth of the estuary during storms. The annual floods of the Nile in Egypt are well known, and the fertile deposit of mud which they leave on the plains. This mud is *stratified*, the thin layer thrown down in one season differing slightly in colour from that of a previous year, and being separable from it, as has been observed in excavations at Cairo, and other places.*

When beds of sand, clay, and marl, containing shells and vegetable matter, are found arranged in the same manner in the interior of the earth, we ascribe to them a similar origin; and the more we examine their characters in minute detail, the more exact do we find the resemblance. Thus, for example, at various heights and depths in the earth, and often far from seas, lakes, and rivers, we meet with layers of rounded pebbles composed of different rocks mingled together. They are like the pebbles formed in the beds of torrents and rivers, which are carried down into the sea wherever these descend from high grounds bordering a coast. There the gravel is spread out by the waves and currents of the ocean over a considerable space; but during seasons of drought the torrents and rivers are nearly dry, and have only

* See Silliman's Amer. Journ. of Sci. vol. xxviii. 1835; also Principles of Geology, *Index*, " Nile," " Rivers," &c.

power to convey fine sand or mud into the sea.
Hence, alternate layers of gravel and fine sediment accumulate under water, and such alternations are found by geologists in the interior of
every continent. *

If a stratified arrangement, and the rounded
forms of pebbles, are alone sufficient to lead
us to the conclusion that certain rocks originated under water, this opinion is farther confirmed by the distinct and independent evidence
of *fossils*, so abundantly included in the earth's
crust. By a *fossil* is meant any body, or the
traces of the existence of any body, whether animal or vegetable, which has been buried in the
earth by natural causes. Now the remains of
animals, especially of aquatic species, are found
almost everywhere imbedded in stratified rocks.
Shells and corals are the most frequent, and with
them are often associated the bones and teeth of
fish, fragments of wood, impressions of leaves, and
other organic substances. Fossil shells of forms
such as now abound in the sea, are met with far
inland, both near the surface and at all depths
below it, as far as the miner can penetrate. They
occur at all heights above the level of the ocean,
having been observed at an elevation of from
8000 to 9000 feet in the Alps and Pyrenees, more

* See Principles of Geology by the author; refer to ' Mag-
nan,' and ' Conglomerates,' in the *Index* of different editions.

than 13,000 feet high in the Andes, and above 15,000 feet in the Himalayas.

These shells belong mostly to marine testacea, but in some places exclusively to forms characteristic of lakes and rivers. Hence we conclude that some ancient strata were deposited at the bottom of the sea, while others were formed in lakes and estuaries.

When geology was first cultivated it was a general belief, that these marine shells and other fossils were the effects and proofs of the general deluge. But all who have carefully investigated the phenomena have long rejected this doctrine. A transient flood might be supposed to leave behind it, here and there upon the surface, scattered heaps of mud, sand, and shingle, with shells confusedly intermixed; but the strata containing fossils are not superficial deposits, and do not cover the earth, but constitute the entire mass of mountains. It has been also the favourite notion of some modern writers, who are aware that fossil bodies cannot all be referred to the deluge, that they, and the strata in which they are entombed, may have been deposited in the bed of the ocean during a period of several thousand years which intervened between the creation of man and the deluge. They imagine that the antediluvian bed of the ocean, after having been the receptacle of many stratified deposits, became converted, at the

time of the flood, into the lands which we inhabit,
and that the ancient continents were at the same
time submerged, and became the bed of the
present sea. This hypothesis, however preferable
to the diluvial theory, as admitting that all
fossiliferous strata were slowly and successively
thrown down from water, is yet wholly inadequate
to explain the repeated revolutions which the earth
has undergone, and the signs which the existing
continents exhibit, in most regions, of having
emerged from the ocean at an era far more remote
than four thousand years from the present time.
It will also be seen in the sequel, that many
distinct sets of sedimentary strata, each several
hundreds or thousands of feet thick, are piled one
upon the other in the earth's crust, each containing their peculiar fossil animals and plants,
which are distinguishable with few exceptions
from species now living. The mass of some of
these strata consists almost entirely of corals, others
are made up of shells, others of plants turned
into coal, while some are without fossils. In one
set of strata the species of fossils are marine,
in another, placed immediately above or below,
they as clearly prove that the deposit was formed
in an estuary or lake. When the student has
more fully examined into these appearances, he
will become convinced that the time required for
the origin of the actual continents must have been

far greater than that which is conceded by the
theory above alluded to, and that no one universal
and sudden conversion of sea into land will account
for geological appearances.

We have now pointed out one great class of
rocks, which, however they may vary in mineral
composition, colour, grain, or other characters,
external and internal, may nevertheless be grouped
together as having a common origin. They
have all been formed under water, in the same
manner as sand, mud, shingle, banks of shells,
coral, and the like, and are characterized by stratification or fossils, or by both.

Volcanic rocks. — The division of rocks which we
may next consider are the volcanic, or those which
have been produced, whether in ancient or modern
times, not by water, but by the action of fire, or
subterranean heat. These rocks are for the most
part unstratified, and are devoid of fossils. They
are more partially distributed than aqueous formations, at least in respect to horizontal extension.
Among those parts of Europe where they exhibit
characters not to be mistaken, I may mention not
only Sicily and the country round Naples, but
Auvergne, Velay, and Vivarais, now the departments of Puy de Dome, Haute Loire, and
Ardèche, towards the centre and south of France,
in which we find several hundred conical hills
having the forms of modern volcanos, with craters

more or less perfect on many of their summits.
These cones are composed moreover of lava, sand,
and ashes, similar to those of active volcanos.
Streams of lava may sometimes be traced proceeding from the cones into the adjoining valleys,
where they choke up the ancient channels of rivers
with solid rock, in the same manner as some
modern flows of lava in Iceland have been known
to do, the rivers either flowing beneath or cutting
out a narrow passage on one side of the lava.
Although none of these French volcanos have been
in activity within the period of history or tradition,
their forms are often very perfect. Some however have been compared to the mere skeletons
of volcanos, the rains and torrents having washed
their sides, and removed all the loose sand and
scoriæ, leaving only the harder and more solid
materials. By this erosion, and by earthquakes,
their internal structure has occasionally been laid
open to view, in fissures and ravines ; and we then
behold not only many successive beds and masses
of porous lava, sand, and scoriæ, but also perpendicular walls, or *dikes*, as they are called, of
volcanic rock, cutting through the other materials.
Such dikes are also observed in the structure of
Vesuvius, Etna, and other active volcanos. They
have been formed by the pouring of melted
matter, whether from above or below, into open
fissures, and they commonly traverse deposits of
volcanic tuff, a substance produced by the shower-

ing down from the air, or incumbent waters, of
sand and cinders, first shot up from the interior
of the earth by explosions of volcanic gases.

Besides the parts of France above alluded
to, there are other countries, as the north of
Spain, the south of Sicily, the Tuscan territory of
Italy, the lower Rhenish provinces, and Hungary,
where spent volcanos may be seen with cones,
craters, and often accompanying lava-streams.

There are also other rocks in England, Scotland, Ireland, and almost every country in Europe,
which we infer to be of igneous origin, although
they do not form hills with cones and craters.
Thus, for example, we feel assured that the rock
of Staffa, and that of the Giants' Causeway,
called basalt, is volcanic, because it agrees in its
columnar structure and mineral composition with
streams of lava which we know to have flowed
from the craters of volcanos. We find also
similar basaltic rocks associated with beds of *tuff*
in various parts of the British Isles, and forming
dikes, such as have been spoken of ; and some of
the strata through which these dikes cut are occasionally altered at the point of contact, as if they
had been exposed to the intense heat of melted
matter.

The absence of cones and craters, and long
narrow streams of lava, in England and elsewhere,
is principally attributed by geologists to the eruptions having been formerly submarine, just as a

considerable proportion of volcanos in our own
times burst out beneath the sea. But this question
must be enlarged upon more fully in the chapters
on Igneous Rocks, in which it will also be shewn,
that as different sedimentary formations, containing each their characteristic fossils, have been deposited at successive periods, so also volcanic sand
and scoriæ have been thrown out, and lavas have
flowed over the land or bed of the sea, at many
different epochs, or have been injected into
fissures ; so that the igneous as well as the aqueous
rocks may be classed as a chronological series of
monuments, throwing light on a succession of
events in the history of the earth.

Plutonic rocks. — We have now therefore pointed
out the existence of two distinct orders of mineral
masses, the aqueous and the volcanic : but if we
examine a large portion of a continent, especially
if it contain within it a lofty mountain range, we
rarely fail to discover two other classes of rocks,
very distinct from either of those above alluded to,
and which we can neither assimilate to deposits
such as are now accumulated in lakes or seas, nor to
those generated by ordinary volcanic action. The
members of both these divisions of rocks agree in
being highly crystalline and destitute of organic
remains. The rocks of one division have been
called plutonic, comprehending all the granites
and certain porphyries, which are nearly allied in

some of their characters to volcanic formations. The members of the other class are stratified and often slaty, and have been called by some the *crystalline schists*. In these are included gneiss, micaceous-schist (or mica-slate), hornblende-schist, statuary marble, the finer kinds of roofing slate, and other rocks afterwards to be described. As it is admitted that nothing strictly analogous to these crystalline productions can now be seen in the progress of formation on the earth's surface, it will naturally be asked, on what data we can find a place for them in a system of classification founded on the origin of rocks. First then, in regard to the plutonic class, a passage has been traced from various kinds of granite into different varieties of rocks decidedly volcanic ; so that if the latter are of igneous origin, it is scarcely possible to refuse to admit that the granites are so likewise. Secondly, large masses of granite are found to send forth dikes and veins into the contiguous strata, very much in the same way as lava and volcanic matter penetrate aqueous deposits, both the massive granite and the veins causing changes analogous to those which lava and volcanic gases are known to produce. But the plutonic rocks differ from the volcanic, not only by their more crystalline texture, but also by the absence of tuffs and breccias, which are the products of eruptions at the earth's surface. They differ also by the absence of pores or cellular cavities, which the entangled gases give rise to in ordinary lava. From these and other peculiarities it has been inferred, that the granites have been formed at great depths in the earth, and have cooled and crystallized slowly under enormous pressure where the contained gases could not expand. The volcanic rocks, on the contrary, although they also have risen up from below, have cooled from a melted state more rapidly upon or near the surface. From this hypothesis of the great depth at which the granites originated, has been derived the name of " Plutonic rocks," which they have received to distinguish them from the volcanic. The beginner will easily conceive that the influence of subterranean heat may extend downwards from the crater of every active volcano to a great depth below, perhaps several miles or leagues (see Frontispiece), and the effects which are produced deep in the bowels of the earth may, or rather must be distinct; so that volcanic and plutonic rocks, each different in texture, and sometimes even in composition, may originate simultaneously, the one at the surface, the other far beneath it.

Although granite has often pierced through other strata, it has rarely, if ever, been observed to rest upon them as if it had overflowed. But as this is continually the case with the volcanic rocks, they have been styled from this peculiarity, " overlying " by Dr. MacCulloch; and Mr. Necker has proposed the term "underlying" for the granites, to designate the opposite mode in which they almost invariably present themselves.

Metamorphic rocks. — The fourth and last great division of rocks are the crystalline strata or schists, called gneiss, mica-schist, clay-slate, chlorite-schist, marble, and the like, the origin of which is more doubtful than that of the other three classes. They contain no pebbles or sand or scoriæ, or angular pieces of imbedded stone, and no traces of organic bodies, and they are often as crystalline as granite, yet are divided into beds, corresponding in form to those of sedimentary formations, and are therefore said to be stratified. The beds sometimes consist of an alternation of substances varying in colour, composition, and thickness, precisely as we see in stratified fossiliferous deposits.   According to the theory which I adopt as most probable, and which will be afterwards more fully explained, the materials of these strata were originally deposited from water in the usual form of sediment, but they were subsequently altered by subterranean heat, so as to assume a new texture.   It is demonstrable, in some cases at least, that such a complete conversion has actually taken place.   I have already remarked that alterations, such as might be produced by intense heat, are observed in strata near their contact with veins and dikes of volcanic rocks. These, however, are on a small scale; but a similar influence has been exerted much more powerfully in the neighbourhood of plutonic rocks under different circumstances, and perhaps in combination with other causes. The effects thereby superinduced on fossiliferous strata have sometimes extended to a distance of a quarter of a mile from the point of contact. Throughout the greater part of this space the fossiliferous beds have exchanged an earthy for a highly crystalline texture, and have lost all traces of organic remains. Thus, for example, dark limestones, replete with shells and corals, are turned into white statuary marble, and hard clays into slates called mica-schist and hornblende-schist, all signs of organic bodies having been obliterated.

Although we are in a great degree ignorant of the precise nature of the influence here exerted, yet it evidently bears some analogy to that which volcanic heat and gases are capable of producing; and the action may be conveniently called plutonic, because it appears to have been developed in those regions where plutonic rocks are generated, and under similar circumstances of pressure and depth in the earth. Whether electricity or any other causes have co-operated with heat to produce this influence, may be matter of speculation, but the plutonic influence has sometimes pervaded entire mountain masses of strata. The phenomena, therefore, being sometimes on so grand a scale, we must not consider that the strata have always assumed their crystalline or altered texture in consequence of the proximity of granite, but rather that granite itself, as well as the altered strata, have derived their crystalline texture from plutonic agency.

In accordance with this hypothesis I have proposed (see Principles of Geology), the term "Metamorphic" for the altered strata, a term derived from μετα, meta, *trans*, and μορφη, morphe, *forma*.

Hence there are are four great classes of rocks considered in reference to their origin, — the aqueous, volcanic, plutonic, and metamorphic, all of which may be conceived to have been formed contemporaneously at every geological period, and to be now in the progress of formation. By referring to the Frontispiece, the reader will perceive what relative positions the members of these four great classes A, B, C, D, may occupy in the earth's crust, while in the course of simultaneous production. Thus, while the aqueous deposits A, which are expressed by the yellow colour, have been accumulating in successive strata at the bottom of the sea, the volcanic cone B, has been piled up during a long series of eruptions, and the other igneous rocks coloured purple have also ascended from below in a fluid state. Some of these last have been poured forth into the sea, and there mingled with aqueous sediment. On pursuing downwards either the small dikes or large masses of volcanic rock, we find them pass gradually into plutonic formations, D, which are coloured red, and which underlie all the rest. These last again are seen to be in contact with a zone of contemporaneous metamorphic strata, C, coloured blue, which they penetrate in numerous veins.

In that part of the section which is uncoloured, a more ancient series of mineral masses are seen, belonging also to the four great divisions of rocks. The strata from a to i represent as many distinct aqueous formations, which have originated at different periods, and are each distinguished by their peculiar fossils. The mass $v\,v$ is of volcanic origin, and was formed at one of those periods, namely, when the strata g were deposited. The strata $m\,m$ are ancient metamorphic formations, and the rocks 1, 2, are plutonic, also ancient, but of different dates.

Now it will be shewn in the course of this volume, that portions of each of these four distinct classes of rocks have originated at many successive periods. It is not true, as was formerly supposed, that all granite, together with the crystalline or metamorphic strata, were first formed, and therefore entitled to be called "primitive," and

that the aqueous and volcanic rocks were afterwards superimposed, and should, therefore, rank
as secondary in the order of time. This idea was
adopted in the infancy of the science, when all
formations, whether stratified or unstratified,
earthy or crystalline, with or without fossils, were
alike regarded as of aqueous origin. At that
period it was naturally argued, that the foundation
must be older than the superstructure. Granite,
as being the lowest rock, must have been first
" precipitated from the waters of the primeval
ocean which originally invested the globe," then
the crystalline, and finally the fossiliferous strata,
together with other associated rocks, were deposited.

But when the doctrine of the igneous origin of
granite was generally adopted, the terms primitive and primary, as embracing the plutonic and
metamorphic rocks, should at once have been
banished from the nomenclature of geology. For
after it had been first proved that granite had
originated at many different epochs, some antecedent, others subsequent to the origin of many
fossiliferous strata, it was also demonstrated that
strata which had once contained fossils, had become metamorphic at different periods ; in other
words, some of the rocks termed primary were
newer than others which were called secondary.
A question, therefore, has arisen, whether the

lower crystalline portions of the earth's crust, partially modified as they have been, and renewed from time to time, are newer or older, regarded as a whole, than the sedimentary and volcanic formations. Have the operations of decay and repair been most active above or below? The same question might be asked with respect to the relative antiquity of the foundations and the buildings in certain ancient cities, such as Venice or Amsterdam, which are supported on wooden piles — whether in the course of ages have the wooden props, or the buildings of brick, stone, and marble which they support, proved the most durable? Which have been renewed most frequently? for the piles, when rotten, can be removed one after the other without injury to the buildings above. In like manner the materials of the lower part of the earth's crust may pass from a solid to a fluid state, and may then again become consolidated; or sedimentary strata may assume a new and metamorphic texture, while the strata above continue unchanged, or retain characters by which their claim to high antiquity may be recognized. During such subterranean mutations, the earthquake may shatter and dislocate the incumbent crust, or the ground may rise or sink slowly and insensibly throughout wide areas *; or there may

* See chap. 5.

be volcanic eruptions here and there; but the great mass may not undergo such an alteration as to be regenerated and composed of new rocks.

As all the crystalline rocks may, in some re-spects, be viewed as belonging to one great family, whether they be stratified or unstratified, it will often be convenient to speak of them by one com-mon name. But the use of the term primary would imply a manifest contradiction, for reasons which the student will now comprehend. It is indispensable, therefore, to find a new name, one which must not be of chronological import, and must express, on the one hand, some peculiarity equally attributable to granite and gneiss (to the plutonic as well as the *altered* rocks), and, on the other, must have reference to characters in which those rocks differ, both from the volcanic and from the *unaltered* sedimentary strata. I have proposed in the Principles of Geology the term " hypogene " for this purpose, derived from ὑπο, *under*, and γινομαι, *to be born;* a word implying the theory that granite, gneiss, and the other crys-talline formations are alike *nether-formed* rocks, or rocks which have not assumed their present form and structure at the surface. It is true that all meta-morphic strata must have been deposited originally at the surface, or on that part of the exterior of the globe which is covered by water; but, accord-ing to the views above set forth, they could never

have acquired their crystalline texture, unless
they had been modified by plutonic agency under
pressure in the depths of the earth.

From what has now been said, the reader will
understand that the four great classes of rocks
may each be studied under two distinct points of
view : first, they may be studied simply as mineral
masses deriving their origin from particular causes,
and having a certain composition, form, and position in the earth's crust, or other characters both
positive and negative, such as the presence or
absence of organic remains.   In the second place,
the rocks of each class may be viewed as a grand
chronological series of monuments, attesting a succession of events in the former history of the globe
and its living inhabitants.

I shall accordingly divide this work into two
parts, in reference to these two modes of considering each family of rocks.   In the first part, the
characters of the aqueous, volcanic, plutonic, and
metamorphic rocks will be described, without
reference to their ages, or the periods when they
were formed.   In the second, their different ages
will be considered, and I shall endeavour to explain the rules according to which the chronology
of rocks in each of the four classes may be determined.

CHAPTER II.

AQUEOUS ROCKS — THEIR COMPOSITION AND FORMS OF STRATIFICATION.

Mineral composition of strata — Arenaceous rocks — Argillaceous — Calcareous — Gypsum — Forms of stratification — Original horizontality — thinning out — Diagonal arrangement — Ripple mark.

First, then, in pursuance of the arrangement explained in the last chapter, we have to examine the aqueous or sedimentary rocks, which are for the most part distinctly stratified, and contain fossils. We are to consider them with reference to their mineral composition, external appearance, position, mode of origin, and other characters which belong to them as aqueous formations, without reference to their age, or the various geological periods when they may have originated.

I have already given an outline of the data which lead to the belief that the stratified and fossiliferous rocks were originally deposited under water; but, before entering into a more detailed investigation, it will be desirable to say something of the ordinary materials of which such strata are composed. These may be said to belong principally to three divisions, the arenaceous, the argillaceous, and the calcareous which are formed

c

respectively of sand, clay, and carbonate of lime.
Of these, the arenaceous, or sandy masses, are
chiefly made up of siliceous or flinty grains; the
argillaceous, or clayey, of a mixture of siliceous
matter, with a certain proportion, about a fourth
in weight, of aluminous earth ; and, lastly, the
calcareous rocks or limestones consist of carbonic
acid and lime.

Arenaceous or siliceous rocks. — To speak first of
the sandy division: beds of loose sand are frequently met with, of which the grains consist entirely of silex, which term comprehends all purely
siliceous minerals, as quartz and common flint.
Quartz is silex in its purest form; flint usually
contains some admixture of alumine and oxide of
iron. The siliceous grains in sand and sandstone
are usually rounded, as if by the action of running
water; but they sometimes, though more rarely,
consist of small crystals, as if they had been chemically precipitated from a fluid containing silex
in solution.

Sandstone is an aggregate of such grains, which
often cohere together without any visible cement,
but more commonly are bound together by a slight
quantity of siliceous or calcareous matter, or by
iron or clay. In nature there is every intermediate
gradation, from perfectly loose sand, to the hard-
est sandstone. In *micaceous sandstones* mica is
abundant; and the thin silvery plates into which

that mineral divides, are arranged in layers
parallel to the planes of stratification, giving a
slaty or laminated texture to the rock.

When sandstone is coarse-grained, it is usually
called *grit.* If the grains are rounded, and large
enough to be called pebbles, it becomes a *conglomerate*, or *pudding-stone*, which may consist of
pieces of one or of many different kinds of rock.
A conglomerate, therefore, is simply gravel bound
together by a cement.

Argillaceous rocks. — Clay, strictly speaking, is a
mixture of silex or flint with a large proportion,
usually about one fourth, of the substance called
alumine, or argil; but, in common language, any
earth which possesses sufficient ductility, when
kneaded up with water, to be fashioned like paste
by the hand, or by the potter's lathe, is called a
clay; and such clays vary greatly in their composition, and are, in general, nothing more than
mud derived from the decomposition or wearing
down of various rocks. The purest clay found in
nature is porcelain clay, or kaolin, which results
from the decomposition of a rock composed of
felspar and quartz, and it is almost always mixed
with quartz.* *Shale* has also the property, like

* The kaolin of China consists of 71·15 parts of silex,
15·86 of alumine, 1·92 of lime, and 6·73 of water, (W. Phillips, Mineralogy, p. 33.); but other porcelain clays differ
materially, that of Cornwall being composed of 60 parts
of alumine and 40 of silex. (Ibid.)

clay, of becoming plastic in water: it is a more
solid form of clay, having been probably condensed by pressure. It usually divides into thin
laminæ.

One general character of all argillaceous rocks
is to give out a peculiar odour when breathed
upon, which is a test of the presence of alumine,
although it does not belong to pure alumine, but,
apparently, to the combination of that substance
with oxide of iron. *

Calcareous rocks.—This division comprehends
those rocks which, like chalk, are composed of
lime and carbonic acid. Shells and corals are
also formed of the same elements, with the addition
of animal matter. To obtain pure lime it is
necessary to calcine these calcareous substances,
that is to say, to expose them to heat of sufficient
intensity to drive off the carbonic acid, and other
volatile matter, without vitrifying or melting the
lime itself. White chalk is often pure carbonate
of lime; and this rock, although usually in a soft
and earthy state, is sometimes sufficiently solid to
be used for building, and even passes into a
compact stone, or a stone of which the separate
parts are so minute as not to be distinguishable
from each other by the naked eye.

Many limestones are made up entirely of minute

* See W. Phillips's Mineralogy, " Alumine."

fragments of shells and coral, or of calcareous sand cemented together. These last might be called " calcareous sandstones; " but that term is more properly applied to a rock in which the grains are partly calcareous and partly siliceous, or to quartzose sandstones, having a cement of carbonate of lime.

The variety of limestone called " oolite " is composed of numerous small egg-like grains, resembling the roe of a fish, each of which has usually a small fragment of sand as a nucleus, around which concentric layers of calcareous matter have accumulated.

Any limestone which is sufficiently hard to take a fine polish is called *marble*. Many of these are fossiliferous; but statuary marble, which is also called saccharine limestone, as having a texture resembling that of loaf-sugar, is devoid of fossils, and a member of the metamorphic series.

Siliceous limestone is an intimate mixture of carbonate of lime and flint, and is harder in proportion as the flinty matter predominates.

The presence of carbonate of lime in a rock may be ascertained by applying to the surface a small drop of diluted sulphuric, nitric, or muriatic acids; for the lime, having a stronger chemical affinity for any one of these acids than for the carbonic, unites itself immediately with them to form new compounds, thereby becoming a sulphate, nitrate, or muriate of lime. The carbonic acid, when thus liberated from its union with the lime, escapes in a gaseous form, and froths up or effervesces as it makes its way in small bubbles through the drop of liquid. This effervescence is brisk or feeble in proportion as the limestone is pure or impure, or, in other words, according to the quantity of foreign matter mixed with the carbonate of lime. Without the aid of this test, the most experienced eye cannot always detect the presence of lime in rocks.

The above-mentioned three classes of rocks, the arenaceous, argillaceous, and calcareous, pass continually into each other, and rarely occur in a perfectly separate and pure form. Thus it is an exception to the general rule to meet with a limestone as pure as ordinary white chalk, or with clay as aluminous as that used in Cornwall for porcelain, or with sand so entirely composed of siliceous grains as the white sand of Alum Bay in the Isle of Wight, or sandstone so pure as the grit of Fontainebleau, used for pavement in France. More commonly we find sand and clay, or clay and marl, intermixed in the same mass. When the sand and clay are each in considerable quantity, the mixture is called *loam*. If there is much calcareous matter in clay it is called *marl;* but this term has unfortunately been used

so vaguely, as often to be very ambiguous. It has
been applied to substances in which there is no
lime; as, to that red loam usually called red marl
in certain parts of England. Agriculturists were
in the habit of calling any soil a marl, which, like
true marl, fell to pieces readily on exposure to the
air. Hence arose the confusion of using this
name for soils which, consisting of loam, were
easily worked by the plough, though devoid of
lime.

Marl slate bears the same relation to marl which
shale bears to clay, being a calcareous shale. It
is very abundant in some countries, as in the
Swiss Alps. Argillaceous or marly limestone is
also of common occurrence.

There are few other kinds of rock which enter
so largely into the composition of sedimentary strata
as to make it necessary to dwell here on their
characters. I may, however, mention two others,
—magnesian limestone or dolomite, and gypsum.
Magnesian limestone is composed of carbonate of
lime and carbonate of magnesia : the proportion of
the latter amounting in some cases to nearly one
half. It effervesces much more slowly and feebly
with acids than common limestone. In England
this rock is generally of a yellowish colour; but it
varies greatly in mineralogical character, passing
from an earthy state to a white compact stone of
great hardness. *Dolomite*, so common in many

c 4

parts of Germany and France, is also a variety
of magnesian limestone, usually of a granular
texture.

Gypsum. — Gypsum is a rock composed of sulphuric acid, lime, and water. It is usually a soft
whitish-yellow rock, with a texture resembling
that of loaf-sugar, but sometimes it is entirely
composed of lenticular crystals. It is insoluble
in acids, and does not effervesce like chalk and
dolomite, the lime being already combined with
sulphuric acid, for which it has a stronger affinity
than for any other. Anhydrous gypsum is a rare
variety, into which water does not enter as a component part. Gypseous marl is a mixture of
gypsum and marl.

Forms of stratification. — A series of strata
sometimes consists of one of the above rocks,
sometimes of two or more in alternating beds.
Thus, in the coal districts of England, for example,
we often pass through several beds of sandstone,
some of finer, others of coarser grain, some white,
others of a dark colour, and below these, layers of
shale and sandstone or beds of shale, divisible into
leaf-like laminæ, and containing beautiful impressions of plants. Then again we meet with beds of
pure and impure coal, alternating with shales, and
underneath the whole, perhaps, are calcareous
strata, or beds of limestone, filled with corals and
marine shells, each bed distinguishable from an-

other by certain fossils, or by the abundance of particular species of shells or zoophytes.

This alternation of different kinds of rock produces the most distinct stratification; and we often find beds of limestone and marl, conglomerate and sandstone, sand and clay, recurring again and again, in nearly regular order, throughout a series of many hundred strata. The causes which may produce these phenomena are various, and have been fully discussed in my treatise on the modern changes of the earth's surface. * It is there seen that rivers flowing into lakes and seas are charged with sediment, varying in quantity, composition, colour, and grain according to the seasons; the waters are sometimes flooded and rapid, at other periods low and feeble, different tributaries, also, draining peculiar countries and soils, and therefore charged with peculiar sediment, are swollen at distinct periods. It was also shewn that the waves of the sea and currents undermine the cliffs during wintry storms, and sweep away the materials into the deep, after which a season of tranquillity succeeds, when nothing but the finest mud is spread by the movements of the ocean over the same submarine area.

It is not the object of the present work to give a description of these operations, repeated as they

* Consult Index to Prin. of Geol. " Stratification," " Currents," Deltas," " Water," &c.

are, year after year, and century after century; but
I may suggest an explanation of the manner in
which some micaceous sandstones have originated,
those in which we see innumerable thin layers of
mica dividing layers of fine quartzose sand. I observed the same arrangement of materials in recent
mud deposited in the estuary of La Roche St. Ber-
nard in Brittany, at the mouth of the Loire. The
surrounding rocks are of gneiss, which, by its
waste, supplies the mud: when this dries at low
water, it is found to consist of brown laminated
clay, divided by thin seams of mica. The separation of the mica in this case, or in that of
micaceous sandstones, may be thus understood.
If we take a handful of quartzose sand, mixed
with mica, and throw it into a clear running
stream, we see the materials immediately sorted
by the water, the grains of quartz falling almost
directly to the bottom, while the plates of mica
take a much longer time to reach the bottom, and
are carried farther down the stream. At the first
instant the water is turbid, but immediately after
the flat surfaces of the plates of mica are seen alone
reflecting a silvery light, and they descend slowly,
to form a distinct micaceous lamina. The mica is
the heavier mineral of the two; but it remains
longer suspended, owing to its great extent of surface. It is easy, therefore, to conceive how the inter-
mittent action of waves, currents, and tides, may

sort the sediment brought down from the waste of a granitic country, and throw down the mica, layer after layer, separately from the mud or sand.

Original horizontality. — It has generally been said that the upper and under surfaces of strata, or the planes of stratification, as they are termed, are parallel. Although this is not strictly true, they make an approach to parallelism, for the same reason that sediment is usually deposited at first in nearly horizontal layers. The reason of this arrangement can by no means be attributed to an original evenness or horizontality in the bed of the sea; for it is ascertained that in those places where no matter has been recently deposited, the bottom of the ocean is often as uneven as that of the dry land, having in like manner its hills, valleys, and ravines. Yet if the sea should sink, or the water be removed near the mouth of a large river where a delta has been forming, we should see extensive plains of mud and sand laid dry, which, to the eye, would appear perfectly level, although, in reality, they would slope gently from the land towards the sea. This tendency in newly-formed strata to assume a horizontal position, arises principally from the motion of the water, which forces along particles of sand or mud at the bottom, and causes them to settle in hollows or depressions, where they are less exposed to the force of a current than when they are resting on elevated points. The velocity

c 6

of the current and the motion of the superficial waves diminishes from the surface downwards, and is least in those depressions where the water is deepest. A good illustration of the principle here alluded to, may be sometimes seen in the neighbourhood of a volcano, when a section, whether natural or artificial, has laid open to view a succession of various-coloured layers of sand and ashes, which have fallen in showers upon uneven ground. Thus let A, B (Fig. 1.) be two ridges, with an intervening valley. These original inequalities of the surface have been gradually effaced by beds of sand and ashes *c d e*, the surface at *e* being quite level. It will be seen that although the materials of the first layers have accommodated themselves in a great degree to the shape of the ground A B, yet each bed is thickest at

Fig. 1.

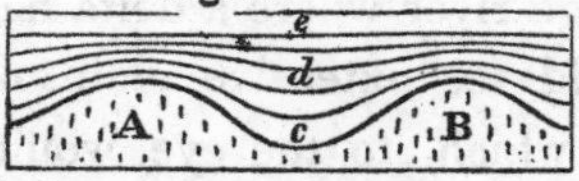

the bottom. At first a great many particles would be carried by their own gravity down the steep sides of A and B, and others would afterwards be blown by the wind as they fell off the ridges, and would settle in the hollow, which would thus become more and more effaced as the strata accumulated from *c* to *e.* This levelling operation may perhaps be rendered more clear to the student by supposing a number of parallel trenches to be dug in a plain of moving sand, like the African desert,

in which case the wind would soon cause all
signs of these trenches to disappear, and the
surface would be as uniform as before. Now,
water in motion can exert this levelling power
on similar materials more easily than air, for
almost all stones lose in water more than a
third of the weight which they have in air, the
specific gravity of rocks being in general as $2\frac{1}{2}$
when compared to that of water, which is estimated at 1. But the buoyancy of sand or mud
would be still greater in the sea, as the density of
salt water exceeds that of fresh.

Yet, however uniform and horizontal may be the
surface of new deposits in general, there are still
many disturbing causes, such as eddies in the water,
and currents moving first in one and then in another
direction, which frequently cause irregularities.
We may sometimes follow a bed of limestone,
shale, or sandstone, for a distance of many hundred yards continuously ; but we generally find at
length that each individual stratum thins out, and

Fig. 2.

Section of strata of sandstone, grit, and conglomerate.

allows the beds which were previously above and
below it to meet. If the materials are coarse, as
in grits and conglomerates, the same beds can

rarely be traced many yards without varying in
size, and often coming to an end abruptly. (See
Fig. 2.)

There is also another phenomenon of frequent
occurrence. We find a series of larger strata, each
of which is composed of a number of minor layers
placed obliquely to the general planes of stratification. To this diagonal arrangement the name
of "false stratification" has been given. Thus in
the annexed section (Fig. 3.) we see seven or eight

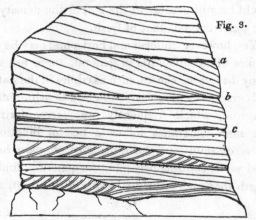

Fig. 3.

Section of sand at Sandy Hill, near Biggleswade, Bedfordshire.
Height twenty feet. (Green-sand formation.)

large beds of loose sand, yellow and brown, and
the lines *a, b, c,* mark some of the principal planes
of stratification, which are nearly horizontal. But
the greater part of the subordinate laminæ do not
conform to these planes, but have often a steep

slope, the inclination being sometimes towards
opposite points of the compass. When the sand is
loose and incoherent, as in the case here represented, the deviation from parallelism of the
slanting laminæ cannot possibly be accounted
for by any rearrangement of the particles acquired
during the consolidation of the rock. In what
manner then can such irregularities be due to
original deposition? We must suppose that at the
bottom of the sea, as well as in the beds of rivers,
the motions of waves, currents, and eddies often
cause mud, sand, and gravel to be thrown down
in heaps on particular spots, instead of being
spread out uniformly over a wide area. Sometimes,
when banks are thus formed, currents may cut
passages through them, just as a river forms its bed.
Suppose the bank A (Fig. 4.) to be thus formed
with a steep sloping side, and the water being

Fig. 4.

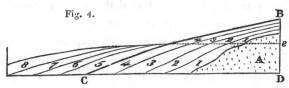

in a tranquil state, the layer of sediment No. 1.
is thrown down upon it, conforming nearly to
its surface. Afterwards the other layers, 2, 3, 4,
may be deposited in succession, so that the bank
B C D is formed. If the current then increases
in velocity, it may cut away the upper portion of

this mass down to the dotted line *e* (Fig. 4.), and
deposit the materials thus removed farther on, so
as to form the layers 5, 6, 7, 8. We have now the
bank B C D E (Fig. 5.), of which the surface is

Fig. 5.

almost level, and on which the nearly horizontal
layers 9, 10, 11, may then accumulate. The opposite slope of the diagonal layers of successive
strata, in the section Fig. 3., may be accounted
for by changes in the direction of the tides and
currents in the same place.

The ripple mark, so common on the surface
of sandstones of all ages (see Fig. 6.), and which
is so often seen on the sea-shore at low tide,
seems to originate in the drifting of materials
along the bottom of the water, in a manner
very similar to that which may explain the inclined layers above described. This ripple is
not entirely confined to the beach between high
and low water mark, but is also produced on
sands which are constantly covered by water.
Similar undulating ridges and furrows may also be
sometimes seen on the surface of drift snow and
blown sand. The following is the manner in
which I once observed the motion of the air to

Fig. 6.

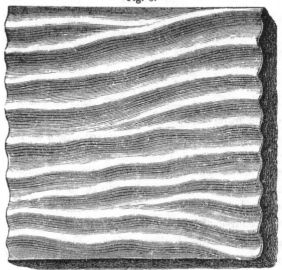

Slab of ripple-marked (new red) sandstone from Cheshire.

produce this effect on a large extent of level
beach, exposed at low tide near Calais. Clouds
of fine white sand were blown from the neighbouring dunes, so as to cover the shore, and
whiten a dark level surface of sandy mud, and
this fresh covering of sand was beautifully
rippled. On levelling all the small ridges
and furrows of this ripple over an area several
yards square, I saw them perfectly restored in
about ten minutes, the general direction of the
ridges being always at right angles to that of the
wind. The restoration began by the appearance

here and there of small detached heaps of sand,
which soon lengthened and joined together, so as
to form long sinuous ridges with intervening
furrows. Each ridge had one side slightly inclined,

Fig. 7.

and the other steep; the lee side being always
steep, as *b*, *c*,—*d*, *e*; the windward side a gentle
slope, as *a*, *b*, —*c*, *d*, Fig. 7. When a gust of wind
blew with sufficient force to drive along a cloud of
sand, all the ridges were seen to be in motion at
once, each encroaching on the furrow before it, and,
in the course of a few minutes, filling the place
which the furrows had occupied. The mode of
advance was by the continual drifting of grains of
sand up the slopes *a b* and *c d*, many of which
grains, when they arrived at *b* and *d*, fell over the
scarps *b c* and *d e*, and were under shelter from
the wind; so that they remained stationary,
resting, according to their shape and momentum,
on different parts of the descent, and a few only
rolling to the bottom. In this manner each ridge
was distinctly seen to move slowly on as often as
the force of the wind augmented. Occasionally
part of a ridge, advancing more rapidly than the
rest, overtook the ridge immediately before it,
and became confounded with it, thus causing those
bifurcations and branches which are so common,

and two of which are seen in the slab Fig. 6.
We may observe this configuration in sandstones
of all ages, and in them also, as now on the sea-
coast, we may often detect two systems of ripples
interfering with each other; one more ancient and
half effaced, and a newer one, in which the grooves
and ridges are more distinct, and in a different
direction. This crossing of two sets of ripples
arises from a change of wind, and the new
direction in which the waves are thrown on the
shore.

CHAPTER III.

HAVING in the last chapter considered the forms
of stratification so far as they are determined by
the arrangement of inorganic matter, we may now
turn our attention to the manner in which organic
remains are distributed through stratified deposits.
We should often be unable to detect any signs of
stratification or of successive deposition, if particular kinds of fossils did not occur here and
there at certain depths in the mass. At one level,
for example, bivalve shells of some one or more
species predominate; at another, some univalve
shell, and at a third, corals; while in some formations we find layers of vegetable matter separating strata.

It may appear inconceivable to a beginner how mountains, several thousand feet thick, can have become filled with fossils from top to bottom ; but the difficulty is removed when he reflects on the origin of stratification, as explained in the last chapter, and allows sufficient time for the accumulation of sediment. He must never lose sight of the fact that, during the process of deposition, each separate layer was once the uppermost, and covered immediately by the water in which aquatic animals lived. Each stratum, in fact, however far it may now lie beneath the surface, was once in the state of loose sand or soft mud at the bottom of the sea, in which shells and other bodies easily became enveloped.

By attending to the nature of these remains, we are often enabled to determine whether the deposition was slow or rapid, whether it took place in a deep or shallow sea, near the shore or far from land, and whether the water was salt, brackish, or fresh. Some limestones consist almost exclusively of corals, and their position has evidently been determined by the manner in which the zoophytes grew ; for if the stratum be horizontal, the round spherical head of certain species is uppermost, and the point of attachment directed downwards. This arrangement is sometimes repeated throughout a great succession of strata. From what we know of the growth of similar zoophytes in modern reefs, we infer that the rate of increase was extremely slow, and some of the fossils must have flourished for ages like forest trees, before they attained so large a size. During these ages, the water remained clear and transparent, for such zoophytes cannot live in turbid water.

In like manner, when we see thousands of full-grown shells dispersed every where throughout a long series of strata, we cannot doubt that time was required for the multiplication of successive generations; and the evidence of slow accumulation is rendered more striking from the proofs, so often discovered, of fossil bodies having lain for a time on the floor of the ocean after death, before they were imbedded in sediment. Nothing, for example, is more common than to see fossil oysters in clay, with serpulæ, acorn-shells, corals, and other creatures, attached to the inside of the valves, so that the mollusk was certainly not buried in argillaceous mud the moment it died. There must have been an interval during which it was still surrounded with clear water, when the testacea, now adhering to it, grew from an embryo state to full maturity. Attached shells which are merely external, like some of the serpulæ in the annexed figure (Fig. 8.), may often have grown upon an oyster or other shell while the animal within was still living; but if they are found on the inside, it could only happen after the death

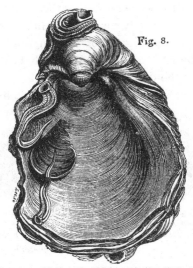

Fig. 8.

Fossil Gryphœa, covered both on the outside and inside with fossil serpulœ.

of the inhabitant of the shell which affords
the support. Thus, in Fig. 8., it will be seen
that two serpulæ have grown on the interior, one
of them exactly on the place where the adductor
muscle of the *Gryphœa* (a kind of oyster) was
fixed.

Some fossil shells, even if simply attached to
the *outside* of others, bear full testimony to the
conclusion above alluded to, namely, that an interval elapsed between the death of the creature
to whose shell they adhere, and the burial of the
same in mud or sand.   The sea-urchins, or *Echini*,
so abundant in white chalk, afford a good illus-

tration.   It is well known that these animals, when
living, are invariably covered with numerous spines,
which serve as organs of motion, and are supported by rows of tubercles, which last are only
seen after the death of the sea-urchin, when the
spines have dropped off.   In Fig. 10. a living spe-

Fig. 9.   Fig. 10.

   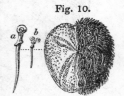

Serpula attached to   Recent Spatangus with the spines
fossil Spatangus   removed from one side.
from the chalk.   b. Spine and tubercles, nat. size.
                         a. The same magnified.

cies of Spatangus, common on our coast, is represented with one half of its shell stripped of the
spines.   In Fig. 9. a fossil of the same genus from
the white chalk of England shews the naked surface which the individuals of this family exhibit
when denuded of their bristles.   The full-grown
Serpula, therefore, which now adheres externally,
could not have begun to grow till the Spatangus
had died, and the spines were detached.

Now the series of events here attested by a single fossil may be carried a step farther.   Thus,
for example, we often meet with a sea-urchin in
the chalk (see Fig. 11.), which has fixed to it the
lower valve of a *crania*, an extinct genus of bivalve

mollusca. The upper valve (*b* Fig. 11.) is almost

Fig. 11.

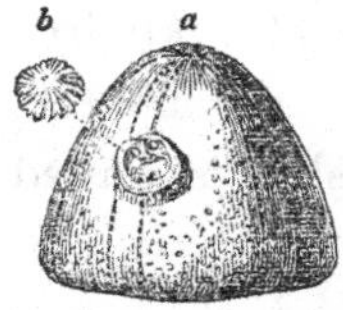

a. Echinus from the chalk, with lower valve of the *Crania* attached.
b. Upper valve of the *Crania* detached.

invariably wanting, though occasionally found in a perfect state of preservation in white chalk at some distance. In this case, we see clearly that the sea-urchin first lived from youth to age, then died and lost its spines, which were carried away. Then the young *Crania* adhered to the bared shell, and perished in its turn; after which the upper valve was separated from the lower before the *Echinus* became enveloped in chalky mud.

It may be well to mention one more illustration of the manner in which single fossils may sometimes throw light on a former state of things, both in the bed of the ocean and on some adjoining land. We meet with many fragments of wood bored by ship-worms at various depths in the clay on which London is built. Entire branches and stems of trees, several feet in length, are sometimes dug out, drilled all over by the holes of these borers, the tubes and shells of the mollusk still remaining in the cylindrical hollows. In Fig. 13., *e*, a representation is given of a piece of recent wood pierced by the *Teredo navalis*, or common ship-worm, which destroys wooden piles and ships. When the cylindrical tube *d* has been extracted from the wood, a shell is seen at the larger ex-

D

tremity, composed of two pieces, as shown at *c.*
In like manner, a piece of fossil wood (*a*, Fig. 12.)
has been perforated by an animal of a kindred
but extinct genus, called *Teredina* by Lamarck.
The calcareous tube of this mollusk was united

Fig. 12.

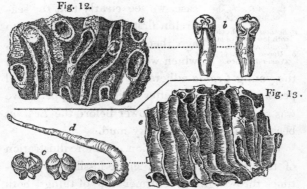

Fig. 13.

Fossil and recent wood drilled by perforating mollusca.

Fig. 12. *a.* Fossil wood from London clay, bored by *Teredina.*
     b. Shell and tube of *Teredina personata*, the right hand
        figure the ventral, the left the dorsal view.
Fig. 13. *e.* Recent wood bored by *Teredo.*
     d. Shell and tube of *Teredo navalis*, from the same.
     c. Anterior and posterior view of the valves of same
        detached from the tube.

and as it were soldered on to the valves of the
shell (*b*), which therefore cannot be detached from
the tube, like the valves of the recent Teredo.
The wood in this fossil specimen is now converted into a stony mass, a mixture of clay
and lime; but it must once have been buoyant
and floating in the sea, when the Teredinæ

lived upon it, perforating it in all directions. Again, before the infant colony settled upon the drift wood, the branch of a tree must have been floated down to the sea by a river, uprooted, perhaps, by a flood, or torn off and cast into the waves by wind; and thus our thoughts are carried back to a prior period, when the tree grew for years on dry land, enjoying a fit soil and climate.

It has been already remarked that there are rocks in the interior of continents, at various depths in the earth, and at great heights above the sea, almost entirely made up of the remains of zoophytes and testacea. Such masses may be compared to modern oyster-beds and coral reefs; and, like them, the rate of increase must have been extremely gradual. But there are a variety of stony deposits in the earth's crust now proved to have been derived from plants and animals of which the organic origin was not suspected until of late years, even by naturalists. Great surprise was therefore created by the recent discovery of Professor Ehrenberg of Berlin, that a certain kind of siliceous stone, called tripoli, was entirely composed of millions of the skeletons or cases of microscopic animalcules. The substance alluded to has long been well known in the arts, being used in the form of powder for polishing stones and metals. It has been procured, among other places,

from Bilin, in Bohemia, where a single stratum,
extending over a wide area, is no less than 14 feet
thick. This stone, when examined with a powerful microscope, is found to consist of the siliceous
cases of infusoria, united together without any

Fig. 14.   Fig. 15.   Fig. 16.

Bacillaria   *Gaillonella*   *Gaillonella*
vulgaris?   *distans.*   *ferruginea.*

*These figures are magnified nearly 300 times, except the lower
figure of G. ferruginea (Fig. 16, a), which is magnified 2000 times.*

visible cement. It is difficult to convey an idea
of their extreme minuteness; but Ehrenberg es-
timates that in the Bilin tripoli there are 41,000
millions of individuals of the *Gaillonella distans*
(see Fig. 15.) in every cubic inch, which weighs
about 220 grains, or about 187 millions in a single
grain. At every stroke, therefore, that we make
with this polishing powder, several millions, perhaps tens of millions, of perfect fossils are crushed
to atoms.

The shells or shields of these infusoria are of
pure silex, and their forms are various, but very
marked and constant in particular genera and
species. Thus, in the family Bacillaria (see
Fig. 14.), the fossil species preserved in tripoli are
seen to exhibit the same divisions and transverse
lines which characterize the living shells of kindred

form. With these, also, the siliceous spieulæ or internal supports of the freshwater sponge, or *Spongilla* of Lamarck, are sometimes intermingled (see the needle-shaped bodies in Fig. 18.). These flinty cases and spiculæ, although hard, are very fragile, breaking like glass, and are therefore admirably adapted, when rubbed, for wearing down into a fine powder fit for polishing the surface of metals.

Fig. 18.   Fig. 17.

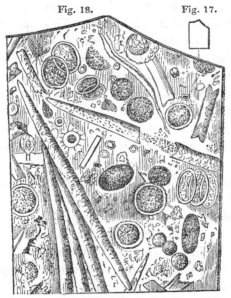

Fragment of semi-opal from the great bed of tripoli, Bilin.

Fig. 17. Natural size.

Fig. 18. The same magnified, showing circular articulations of a species of *Gaillonella*, and spiculæ of *Spongilla*.

Besides the tripoli, which is formed exclusively of infusoria, there occurs in the upper part of the great stratum at Bilin another heavier and more compact stone, a kind of semi-opal, in which innumerable parts of infusoria and spiculæ of the Spongilla are filled with, and cemented together by, siliceous matter. It is supposed that the shells of the more delicate animalcules have been dissolved by water, and have thus given rise to this opal, in which the more durable fossils are preserved like insects in amber. This opinion is confirmed by the fact that the small shells decrease in number and sharpness of outline in proportion as the opaline cement increases in quantity.

In the Bohemian tripoli above described, as in that of Planitz in Saxony, the species of infusoria are freshwater; but in other countries, as in the tripoli of the Isle of France, they are of marine species, and they all belong to formations of the *tertiary* period, which will be spoken of hereafter. (See Part II.)

A well-known substance, called bog-iron ore, often met with in peat-mosses, has also been shown by Ehrenberg to consist of innumerable articulated threads, of a yellow ochre colour, composed partly of flint and partly of oxide of iron. These threads are the cases of a minute animalcule, called *Gaillonella ferruginea* (Fig. 16.).

It is clear that much time must have been re-

quired for the accumulation of strata to which
countless generations of infusoria have contributed
their shells; and these discoveries lead us naturally to suspect that other deposits, of which the
materials have usually been supposed to be
inorganic, may in reality have been derived from
microscopic organic bodies. That this is the case
with the white chalk, has often been imagined,
this rock having been observed to abound in a
variety of marine fossils, such as shells, echini,
corals, sponges, crustacea, and fishes. Mr. Lonsdale, on examining lately, in the museum of the
Geological Society of London, portions of white
chalk from different parts of England, found, on
carefully pulverizing them in water, that what
appear to the eye simply as white grains were, in
fact, well-preserved fossils. He obtained about a
thousand of these from each pound weight of chalk,
some being fragments of minute corallines, others
entire Foraminifera and Cytherinæ. The annexed
drawings will give an idea of the beautiful forms

Cytherinæ and Foraminifera from the chalk.

Fig. 19.   Fig. 20.   Fig. 21.   Fig. 22.

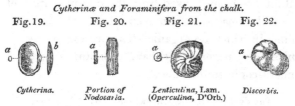

Cytherina.   *Portion of*   *Lenticulina,* Lam.   *Discorbis.*
               Nodosaria.   (*Operculina,* D'Orb.)

of many of these bodies. The figures *a a* represent their natural size, but, minute as they seem,

the smallest of them, such as *a*, Fig. 22., are
gigantic in comparison with the cases of infusoria
before mentioned. There is, moreover, good reason
to believe that the chambers into which these
Foraminifera are divided are actually often filled
with hundreds of infusoria; for many of the minute
grains which they contain, and which compose the
enveloping chalk, have been observed, under a
powerful microscope, to consist of circular discs,
like the articulations of *Gaillonella*, before represented in Fig. 18. The bodies alluded to were
calcareous; but Ehrenberg has discovered others
in the flints of the chalk, which, like the infusoria
in tripoli, are siliceous. These forms are especially
apparent in the white coating of flints, often accompanied by innumerable needle-shaped spiculæ
of sponges: and the same are occasionally visible
in the central parts of chalk flints where they
are of a lighter colour. After reflecting on these
discoveries, we are naturally led on to conjecture
that, as the formless cement in the semi-opal of
Bilin has been derived from the decomposition of
animal remains, so also even those parts of chalk
flints in which no organic structure can be recognized may nevertheless have constituted a part of
microscopic animalcules.

" The dust we tread upon was once alive ! " — *Byron.*

How faint an idea does this exclamation of the

poet convey of the real wonders of nature ! for here we discover proofs that the calcareous and siliceous dust of which hills are composed has not only been once alive, but almost every particle, albeit invisible to the naked eye, still retains the organic structure which, at periods of time incalculably remote, was impressed upon it by the powers of life.

As I have dwelt upon the proofs of the slowness with which fossiliferous strata in general have been produced, I may remark that some writers have argued, from the appearances of certain deposits containing coal, that sedimentary rocks of great thickness have been accumulated with rapidity. This conclusion has been drawn chiefly from a remarkable phenomenon, — the position of the trunks of fossil trees intersecting obliquely, and often at right angles, the planes of many strata. For a full examination of this question, the reader is referred to the chapter on the carboniferous formations, in the sequel ; and I shall merely say here, that, although partial deposits have been thrown down in the spots where these fossil trees occur in a comparatively short lapse of time, yet we can by no means infer that a similar rate of increase of carboniferous rocks prevailed simultaneously over a wide area. On the other hand, the vegetable origin of coal is now universally admitted by geologists ; and, when we discuss the

probable manner in which the terrestrial plants
from which it was derived were imbedded in
marine shale and sandstone, we shall find it necessary to suppose a long succession of operations.

Freshwater and marine fossils. — Strata, whether
deposited in salt or fresh water, have the same
forms; but the fossils are very different in the two
cases, for the same reason that aquatic animals
which frequent lakes and rivers are distinct from
those inhabiting the sea. As an example of
English strata characterized by freshwater fossils,
I may point out a formation which extends over
the northern part of the Isle of Wight, composed
of marl and limestone more than fifty feet thick.
The shells are principally, if not all, of extinct
species ; but they are of the same genera as those
now abounding in ponds and lakes, either in our
own country or warmer latitudes.

In many parts of France, as in Auvergne, for
example, strata of limestone, marl, and sandstone
occur, hundreds of feet thick, which contain exclusively freshwater and land shells, together with
the remains of terrestrial quadrupeds. The number of land shells scattered through some of these
freshwater deposits is exceedingly great; and there
are even districts where the rocks scarcely contain
any other fossils except snail-shells (*helices*) ; as, for
instance, the limestone on the left bank of the
Rhine, between Mayence and Worms, at Oppen-

heim, Findheim, Budenheim, and other places.
In order to account for this phenomenon, the
geologist has only to examine the small deltas of
torrents which enter the Swiss lakes when the
waters are low, such as the newly-formed plain
where the Kander enters the Lake of Thun. He
there sees sand and mud strewed over with innumerable dead land shells, which have been brought
down from valleys in the Alps in the preceding
spring, during the melting of the snows. Again,
if we search the sands on the borders of the Rhine,
in the lower part of its course, we find countless
land shells mixed with others of species belonging
to lakes, stagnant pools, and marshes. These
individuals have been washed away from the
alluvial plains of the great river and its tributaries,
some from mountainous regions, others from the
low country.

Although freshwater formations are often of
great thickness, yet they are usually very limited
in area when compared to marine deposits, just
as lakes and estuaries are of small dimensions in
comparison with seas.

We may distinguish a freshwater formation,
first, by the absence of many fossils almost invariably met with in marine strata. For example,
there are no corals, no sea-urchins, and scarcely
any other zoophytes; no chambered shells, such
as the nautilus, nor microscropic Foraminifera.

But it is chiefly by attending to the forms of the mollusca that we are guided in determining the point in question. In a freshwater deposit, the number of individual shells is often as great, if not greater, than in a marine stratum; but there are fewer species and genera. This might be anticipated from the fact that the genera and species of recent freshwater and land shells are few when contrasted with the marine. Thus, the genera of true mollusca according to Blainville's system, excluding those of extinct species and those without shells, amount to about 200 in number, of which the terrestrial and freshwater genera scarcely form more than a sixth. *

Almost all bivalve shells, or those of acephalous mollusca, are marine, about ten only out of ninety genera being freshwater. Among these last, the four most common forms, both recent and fossil, are Cyclas, Cyrena, Unio, and Anodonta (see figures); the two first and two last of which are so nearly allied as to pass into each other.

Fig. 23.   Fig. 24.

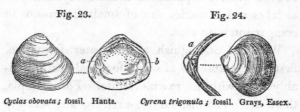

Cyclas obovata; fossil. Hants.   *Cyrena trigonula;* fossil. Grays, Essex.

* See Synoptic Table in Blainville's Malacologie.

Fig. 25.

Fig. 26.

Fig. 27.

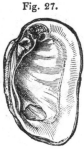

Anodonta Cordierii ;
fossil. Paris.

*Anodonta latimargi-
natus ;* recent. Bahia.

Unio littoralis ;
recent. Auvergne.

Lamarck divided the bivalve mollusca into the *Dimyary*, or those having two large muscular impressions in each valve, as *a b* in the Cyclas, Fig. 23.,

Fig. 28.

*Ostrea vesicularis (Gryphæa
globosa,* Sow.) ; chalk.

and the *Monomyary*, such as the oyster and scallop, in which there is only one of these impressions, as is seen in Fig. 28. Now, as none of these last, or the unimuscular bivalves, are freshwater, we may at once presume a deposit in which we find any of them to be marine.

The univalve shells most characteristic of freshwater deposits are, Planorbis, Limnea, and Paludina. (See figures.) But to these are occasionally added Physa, Succinea, Ancylus, Valvata, Melanopsis, Melania, and Neritina. (See figures.)

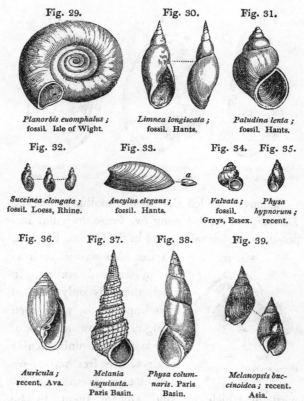

Fig. 29.

Planorbis euomphalus ;
fossil. Isle of Wight.

Fig. 30.

Limnea longiscata ;
fossil. Hants.

Fig. 31.

Paludina lenta ;
fossil. Hants.

Fig. 32.

Succinea elongata ;
fossil. Loess, Rhine.

Fig. 33.

Ancylus elegans ;
fossil. Hants.

Fig. 34.

Valvata ;
fossil.
Grays, Essex.

Fig. 35.

*Physa
hypnorum ;*
recent.

Fig. 36.

Auricula ;
recent, Ava.

Fig. 37.

*Melania
inquinata.*
Paris Basin.

Fig. 38.

*Physa colum-
naris.* Paris
Basin.

Fig. 39.

*Melanopsis buc-
cinoidea ;* recent.
Asia.

In regard to one of these, the Ancylus (Fig. 33.), Mr. Gray observes that it sometimes differs in no respect from the marine Siphonaria, except in the animal. The shell, however, of the Ancylus is usually thinner. *

* Gray, Phil. Trans., 1835, p. 302.

Some naturalists include Neritina (Fig. 40.) and the marine Nerita (Fig. 41.) in the same genus, it being scarcely possible to distinguish the two by good generic characters. But, as a general rule, the fluviatile species are smaller, smoother,

Fig. 40.   Fig. 41.

Neritina globulus. Paris basin,   *Nerita granulosa.* Paris basin.

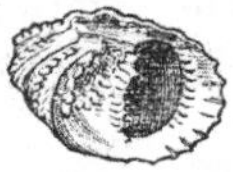

and more globular than the marine; and they have never, like the Neritæ, the inner margin of the outer lip toothed or crenulated. (See Fig. 41.)

A few genera, among which Cerithium (Fig. 42.) is the most abundant, are common both to rivers and the sea, having species peculiar to each. Other genera, like Auricula (Fig. 36.), are amphibious, living both in freshwater and on land.

The terrestrial shells are all univalves. The most abundant genera among these, both in a recent and fossil state, are Helix (Fig. 45.), Cyclostoma, Pupa, (Fig. 44.), Clausilia, Bulimus (Fig. 43.), and Achatina; which two last are nearly allied and pass into each other. The same may be said with almost equal truth of Pupa and Clausilia.

Fig. 42.

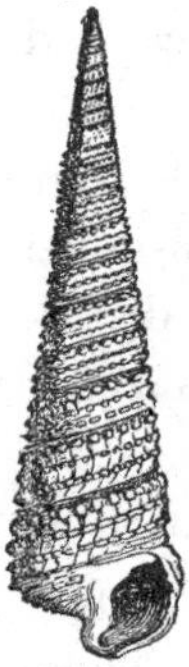

Cerithium cinctum.
Paris basin.

Fig. 43.   Fig. 44.   Fig. 45.

Bulimus   Pupa muscorum.   Helix plebeium.
lubricus.

All recent; and also fossil from Loess of Rhine.

The Ampullaria (Fig. 46.) is another genus of

Fig. 46.

shells, inhabiting rivers and ponds
in hot countries. Many fossil
species have been referred to this
genus, but they have been found
chiefly in marine formations, and
are suspected by some conchologists to belong to Natica and

Ampullaria glauca,
from the Jumna.

other marine genera.

All univalve shells of land and freshwater
species have entire mouths; and this circumstance
may often serve as a convenient rule for distinguishing freshwater from marine strata; since, if
any univalves occur of which the mouths are not
entire, we may conclude that the formation is
marine. The aperture is said to be entire in such
shells as the Ampullaria and the land shells
figured in this page, when its outline is not inter-
rupted by an indentation or notch such as that in
Ancillaria (Fig. 48.); or is not prolonged into a
canal, as that seen at *a* in Pleuromota (Fig. 47.).

The mouths of a large proportion of the marine
univalves have either these notches or canals, and

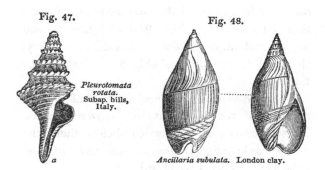

Fig. 47.

Fig. 48.

Pleurotomata rotata. Subap. hills, Italy.

a

Ancillaria subulata. London clay.

all these species are, without exception, carnivorous; whereas nearly all testacea having entire mouths, are plant-eaters, whether the species be marine, freshwater, or terrestrial.

There is, however, one genus which affords an occasional exception to one of the above rules. The Cerithium (Fig. 42.), although provided with a short canal, comprises some species which inhabit salt, others brackish, and others fresh water.

Among the fossils very common in freshwater deposits, are the shells of Cypris, a minute crustaceous animal, having a shell much resembling that of the bivalve mollusca.* Many minute living species of this genus swarm in lakes and stagnant pools in Great Britain; but their shells are not, if considered separately, conclusive as to the freshwater origin of a deposit, because an-

* See figures in chap. on Wealden, Part II.

other kindred genus of the same order, the Cytherina of Lamarck (see Fig. 19. p. 55.), inhabits salt water; and, although the animal differs slightly, the shell is undistinguishable from that of the Cypris.

The seed-vessels of Chara, a genus of aquatic plants, are very frequent in freshwater strata. These seed-vessels were called, before their true nature was known, gyrogonites, and were supposed to be shells. (See Fig. 49. *a*.)

Fig. 49. Fig. 50.

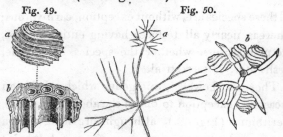

Chara medicaginula ;
fossil. Isle of Wight.
  a, Seed-vessel,
    magnified 20
    diameters.
b, Stem, magnified.

Chara elastica ; recent. Italy.
a, Sessile seed-vessel between the division of the leaves of the female plant.
b, Transverse section of a branch, with five seed-vessels magnified, seen from below upwards.

The Charæ inhabit the bottom of lakes and ponds, and flourish mostly where the water is charged with carbonate of lime. Their seed-vessels are covered with a very tough integument, capable of resisting decomposition; to which circumstance we may attribute their abundance in a fossil state. The annexed figure (Fig. 50.) represents a branch of one of many new species found by Professor Amici in the lakes of northern Italy. The seed-vessel in this plant is more globular than in the British Charæ, and therefore more nearly resembles in form the extinct fossil species found in England, France, and other countries. The stems, as well as the seed-vessels, of these plants are found both in modern shell marl and in ancient freshwater formations. They are generally composed of a large tube surrounded by smaller tubes; the whole stem being divided at certain intervals by transverse partitions or joints. (See *b*, Fig. 49.)

It is not uncommon to meet with layers of vegetable matter, impressions of leaves, and branches of trees, in strata containing freshwater shells; and we also find occasionally the teeth and bones of land quadrupeds, of species now unknown. The manner by which such remains are occasionally carried by rivers into lakes, especially during floods, has been fully treated of in the " Principles of Geology." *

The remains of fish are occasionally useful in determining the freshwater origin of strata. Certain genera, such as carp, perch, pike, and loach, (*Cyprinus*, *Perca*, *Esox*, and *Cobitis*), as also *Lebias*, being peculiar to freshwater. Other ge-

* See Index, " Fossilization."

nera contain some freshwater and some marine species, as *Cottus*, *Mugil*, and *Anguilla*, or eel. The rest are either common to rivers and the sea, as the salmon; or are exclusively characteristic of salt water. The above observations respecting fossil fishes are applicable only to the more modern or tertiary deposits; for in the more ancient rocks the forms depart so widely from those of existing fishes, that it is very difficult, at least in the present state of science, to derive any information from icthyolites, respecting the element in which strata were deposited.

The alternation of marine and freshwater formations both on a small and large scale, are facts well ascertained in geology. When it occurs on a small scale, it may have arisen from the alternate occupation of certain spaces by river-water and the sea; for in the flood season the river forces back the ocean and freshens it over a large area, depositing at the same time its sediment; after which the salt water again returns, and, on resuming its former place, brings with it sand, mud, and marine shells.

There are also lagoons at the mouths of many rivers, as the Nile and Mississippi, which are divided off by bars of sand from the sea, and which are filled with salt and fresh water by turns. They often communicate exclusively with the river for months, years, or even centuries; and then a breach being made in the bar of sand, they are for long periods filled with salt water.

The Lym-Fiord in Jutland offers an excellent illustration of analogous changes ; for, in the course of the last thousand years, the western extremity of this long frith, which is 120 miles in length, including its windings, has been four times fresh and four times salt, a bar of sand between it and the ocean having been as often formed and removed. The last irruption of salt water happened in 1824, when the North Sea entered, killing all the freshwater shells, fish, and plants ; and from that time to the present, the sea-weed *Fucus vesiculosus*, together with oysters and other marine mollusca, have succeeded the Cyclas, Limnea, Paludina, and Charæ. *

But changes like these in the Lym-Fiord, and those before-mentioned as occurring at the mouths of great rivers, will only account for some cases of marine deposits resting on freshwater strata. When we find, as in the south-east of England, a great series of freshwater beds, resting upon one marine formation of great thickness, and again covered by another more than 1000 feet thick, we shall find it necessary to seek for a different explanation of the phenomena. †

* See Principles, Index, " Lym-Fiord."
† See account of Wealden, Part II.

CHAPTER IV.

CONSOLIDATION OF STRATA AND PETRIFACTION OF
FOSSILS.

Chemical and mechanical deposits — Cementing together of
particles — Hardening by exposure to air — Concretionary
nodules — Consolidating effects of pressure — Minerali-
sation of organic remains — Impressions and casts how
formed — Fossil wood — Göppert's experiments — Pre-
cipitation of stony matter most rapid where putrefaction
is going on — Source of lime in solution — Silex derived from decomposition of felspar — Proofs of the
lapidification of some fossils soon after burial, of others
when much decayed.

HAVING spoken in the preceding chapters of the
forms of stratification, both as dependent on the
deposition of inorganic matter and the distribution
of fossils, I may next treat of the consolidation of
stratified rocks, and the petrifaction of imbedded
organic remains.

Chemical and mechanical deposits. — A distinction has been made by geologists between deposits
of a chemical, and those of a mechanical, origin.
By the latter name are designated beds of mud,
sand, or pebbles produced by the action of running water, also accumulations of stones and
scoriæ thrown out by a volcano, which have fallen
into their present place by the force of gravitation.

But the matter which forms a chemical deposit has not been mechanically suspended in water, but in a state of solution until separated by chemical action. In this manner carbonate of lime is often thrown to the bottom of lakes and seas in a solid form, as may be well seen in many parts of Italy, where mineral springs abound, and where the calcareous stone, called travertin, is deposited. In these springs the lime is usually held in solution by an excess of carbonic acid, or by heat if it be a hot spring, until the water, on issuing from the earth, cools or loses part of its acid. The calcareous matter then falls down in a solid state, encrusting shells, fragments of wood and leaves, and binding them together.*

In coral reefs, large masses of limestone are formed by the stony skeletons of zoophytes; and these, together with shells, become cemented together by carbonate of lime, part of which is probably furnished to the sea-water by the decomposition of dead corals. Even shells of which the animals are still living, on these reefs, are very commonly found to be encrusted over with a hard coating of limestone. †

If sand and pebbles are carried by a river into the sea, and these are bound together immediately by carbonate of lime, the deposit may be described

* See Principles, Index, " Calcareous Springs," &c.
† Ibid. " Travertin," " Coral reefs," &c.

as of a mixed origin, partly chemical, and partly mechanical.

Now, the remarks already made in Chapter II. on the original horizontality of strata are strictly applicable to mechanical deposits, and only partially to those of a mixed nature. Such as are purely chemical may be formed on a very steep slope, or may even encrust the vertical walls of a fissure, and be of equal thickness throughout; but such deposits are of small extent, and for the most part confined to vein-stones.

Cementing of particles. — It is chiefly in the case of calcareous rocks that solidification takes place at the time of deposition. But there are many deposits in which a cementing process comes into operation long subsequently. We may sometimes observe, where the water of ferruginous or calcareous springs has flowed through a bed of sand or gravel, that iron or carbonate of lime has been deposited in the interstices between the grains or pebbles, so that in certain places the whole has been bound together into a stone, the same set of strata remaining in other parts loose and incoherent.

Proofs of a similar cementing action are seen in a rock at Kelloway in Wiltshire. A peculiar band of sandy strata, belonging to the group called Oolite by geologists, may be traced through several counties, the sand being for the most part loose and unconsolidated, but becoming stony near

Kelloway. In this district there are numerous fossil shells which have decomposed, having for the most part left only their casts. The calcareous matter hence derived has evidently served, at some former period, as a cement to the siliceous grains of sand, and thus a solid sandstone has been produced. If we take fragments of many other argillaceous grits, retaining the casts of shells, and plunge them into dilute muriatic or other acid, we see them immediately changed into common sand and mud; the cement of lime, derived from the shells, having been dissolved by the acid.

Traces of impressions and casts are often extremely faint. In some loose sands of recent date we meet with shells in so advanced a stage of decomposition as to crumble into powder when touched. It is clear that water percolating such strata may soon remove the calcareous matter of the shell; and, unless circumstances cause the carbonate of lime to be again deposited, the grains of sand will not be cemented together; in which case no memorial of the fossil will remain. The absence of organic remains from many aqueous rocks may be thus explained.

In some conglomerates, like the puddingstone of Hertfordshire, flinty pebbles and sand are united by a siliceous cement so firmly, that if a block be fractured the rent passes as readily through the pebbles as through the cement.

It is probable that many strata became solid at the time when they emerged from the waters in which they were deposited, and when they first formed a part of the dry land. A well-known fact seems to confirm this idea; by far the greater number of the stones used for building and road-making are much softer when first taken from the quarry than after they have been long exposed to the air. Hence it is found desirable to shape the stones which are to be used in architecture while they are yet soft and wet, and while they contain their "quarry-water," as it is called; also to break up stone intended for roads when soft, and then leave it to dry in the air for months that it may harden. Such induration may perhaps be accounted for by supposing the water, which penetrates the minutest pores of rocks, to deposit on evaporation carbonate of lime, iron, silex, and other minerals previously held in solution. These particles, on crystallizing, would not only be deprived themselves of freedom of motion, but would also bind together other portions of the rock which before were loosely aggregated. On the same principle wet sand and mud become as hard as stone when frozen; because one ingredient of the mass, namely, the water, has crystallized, so as to hold firmly together all the separate particles of which the loose mud and sand were composed.

Dr. MacCulloch mentions a sandstone in Sky,

which may be moulded like dough when first found; and another from China, which is compressible by the hand when immersed in water. But it is not merely these compounds which readily admit water to penetrate into them; some simple minerals, says the same writer, which are rigid and as hard as glass in our cabinets, are often flexible and soft in their native beds; this is the case with asbestos, sahlite, tremolite, and calcedony, and it is reported also to happen in the case of the beryl. *

The marl recently deposited at the bottom of Lake Superior, in North America, is soft, and often filled with fresh-water shells; but if a piece be taken up and dried, it becomes so hard that it can only be broken by a smart blow of the hammer. If the lake therefore was drained, such a deposit would be found to consist of strata of marlstone, like that observed in many ancient European formations, and like them containing fresh-water shells.†

It is probable that some of the heterogeneous materials which rivers transport to the sea may at once set under water, like the artificial mixture called pozzolana, which consists of fine volcanic sand charged with about 20 per cent. of iron, and the addition of a small quantity of lime. This substance hardens and becomes a solid stone in

* Dr. MacCulloch, Syst. of Geol. vol. i. p. 123.
† Princ. of Geol., Index, " Superior, Lake."

water, and was used by the Romans in constructing
the foundations of buildings in the sea.

Consolidation in these cases is brought about
by the action of chemical affinity on finely comminuted matter previously suspended in water. After
deposition similar particles seem to exert a mutual
attraction on each other, and congregate together
in particular spots, forming lumps, nodules, and
concretions. Thus in many argillaceous deposits
there are calcareous balls, or spherical concretions,
ranged in layers parallel to the general stratification; an arrangement which took place after the
shale or marl had been thrown down in successive
laminæ; for these laminæ are often traced in the
concretions, remaining parallel to those of the sur-

Fig. 51.

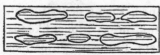

Calcareous nodules in Lias.

rounding unconsoli-
dated rock. (See Fig.
51.) Such nodules
of limestone have
often a shell or other
foreign body in the centre. *

Among the most remarkable examples of concretionary structure are those described by Professor
Sedgwick as abounding in the magnesian limestone of the north of England. The spherical balls
are of various sizes, from that of a pea to a diameter
of several feet, and they have both a concentric and

* See De la Beche's Geological Researches, p. 95.

radiated structure, while at the same time the laminæ of original deposition pass uninterruptedly through them.   In some cliffs this limestone resembles a great irregular pile of cannon balls. Some of the globular masses have their centre in one stratum, while a portion of their exterior passes through to the stratum above or below. Thus the larger spheroid in the annexed section

Fig. 52.

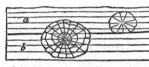

Spheroidal concretions in magnesian limestone.

(Fig. 52.) passes from the stratum *b* upwards into *a*. In this instance we must suppose the deposition of a series of minor layers, first forming the stratum *b*, and afterwards the incumbent stratum *a ;* then a movement of the particles took place, and the carbonates of lime and magnesia separated from the more impure and mixed matter forming the still unconsolidated parts of the stratum.   Crystallization, beginning at the centre, must have gone on forming concentric coats, around the original nucleus without interfering with the laminated structure of the rock.   As to the radiations from a centre it is a phenomenon which, however singular, is common in spherical concretions of various mineral ingredients.

When the particles of rocks have been thus re-arranged by chemical forces, it is sometimes

difficult or impossible to ascertain whether certain
lines of division are due to original deposition or
to the subsequent aggregation of similar particles.

Fig. 53.

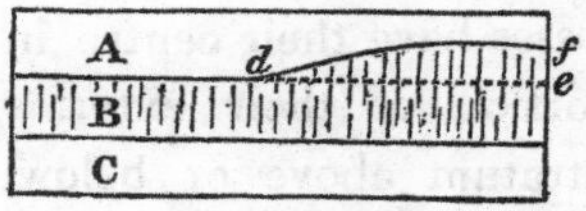

Thus suppose three
strata of grit, A, B, C,
are charged unequally
with calcareous matter,
and that B is the most calcareous. If consolidation takes place in B, the concretionary action may
spread upwards into a part of A, where the carbonate of lime is more abundant than in the
rest; so that a mass, *d, e, f,* forming a portion of
the superior stratum, becomes united with B into
one solid mass of stone. The original line of
division, *d, e,* being thus effaced, the line, *d, f,*
would generally be considered as the surface of
the bed B, though not strictly a true plane of
stratification.

Pressure and heat.—When sand and mud sink to
the bottom of a deep sea the particles are not
pressed down by the enormous weight of the incumbent ocean; for the water, which becomes
mingled with the sand and mud, resists pressure
with a force equal to that of the column of fluid
above. The same happens in regard to organic
remains which are filled with water under great
pressure as they sink, otherwise they would be
immediately crushed to pieces and flattened.
Nevertheless, if the materials of a stratum remain

in a yielding state, and do not set or solidify, they will be gradually squeezed down by the weight of other materials successively heaped upon them, just as soft clay or loose sand on which a house is built may give way.   By such downward pressure particles of clay, sand, and marl may become packed into a smaller space, and be made to cohere together permanently.

Analogous effects of condensation may arise when the solid parts of the earth's crust are forced in various directions by those mechanical movements afterwards to be described, by which strata have been bent, broken, and raised above the level of the sea. Rocks of more yielding materials must often have been forced against others previously consolidated, and, thus compressed, may have acquired a new structure.

But the action of heat at various depths in the earth is probably the most powerful of all causes in hardening sedimentary strata.   To this subject I shall refer again when treating of the metamorphic rocks, and of the slaty and jointed structure.

Mineralization of organic remains.—The changes which fossil organic bodies have undergone since they were first imbedded in rocks, throw much light on the consolidation of strata.   Fossil shells in some modern deposits have been scarcely altered in the course of centuries, having simply

lost a part of their animal matter. But in other
cases the shell has disappeared, and left an impression only of its exterior, or a cast of its interior
form, or, thirdly, a cast of the shell itself, the
original matter of which has been removed.
These different forms of fossilization may easily be
understood if we examine the mud recently thrown
out from a pond or canal in which there are shells.
If the mud be argillaceous, it acquires consistency
on drying, and on breaking open a portion of it
we find that each shell has left impressions of its
external form. If we then remove the shell itself,
we find within a solid nucleus of clay, having the
form of the interior of the shell. This form is
often very different from that of the outer shell.
Thus a cast such as *a*, Fig. 54., commonly called
a fossil screw, would never be suspected by an

Fig. 54.   Fig. 55.

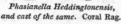

Phasianella Heddingtonensis,
and cast of the same. Coral Rag.

Trochus Anglicus and
cast. Lias.

inexperienced conchologist to be the internal shape
of the fossil univalve, *b*, Fig. 54. Nor should we
have imagined at first sight that the shell *a* and

the cast *b*, Fig. 55., were different parts of the same fossil.   The reader will observe, in the last-mentioned figure (*b*, Fig. 55.), that an empty space shaded dark, which the *shell itself* once occupied, now intervenes between the enveloping stone and the cast of the smooth interior of the whorls.   In such cases the shell has been dissolved and the component particles removed by water percolating the rock.   If the nucleus were taken out a hollow mould would remain, on which the external form of the shell with its tubercles and striæ, as seen in *a*, Fig. 55., would be seen embossed.   Now if the space alluded to between the nucleus and the impression, instead of being left empty, has been filled up with calcareous spar, pyrites, or other mineral, we then obtain from the mould an exact cast both of the external and internal form of the original shell.   In this manner silicified casts of shells have been formed; and if the mud or sand of the nucleus happen to be incoherent, or soluble in acid, we can then procure in flint an empty shell which is the exact counterpart of the original. This cast may be compared to a bronze statue, representing merely the superficial form, and not the internal organization; but there is another description of petrifaction by no means uncommon, and of a much more wonderful kind, which may be compared to certain anatomical models in wax, where not only the outward forms and features, but the nerves, blood-vessels, and other internal organs are also shown. Thus we find corals, originally calcareous, in which not only the general shape, but also the minute and complicated internal organization are retained in flint.

Such a process of petrifaction is still more remarkably exhibited in fossil wood, in which we often perceive not only the rings of annual growth, but all the minute vessels and medullary rays. Many of the minute pores and fibres of plants, and even those spiral vessels which in the living vegetable can only be discovered by the microscope, are preserved. Among many instances I may mention a fossil tree, seventy-two feet in length, found at Gosforth near Newcastle, in sandstone strata associated with coal. By cutting a transverse slice so thin as to transmit light, and magnifying it about fifty-five times, the texture seen in Fig. 56. is exhibited. A texture equally minute and complicated has been observed in the

Fig. 56.

Texture of a tree from the coal-strata, magnified. (Witham.)

wood of large trunks of fossil trees found in the Craigleith quarry near Edinburgh, where the stone was not in the slightest degree siliceous, but consisted chiefly of carbonate of lime, with oxide of iron, alumina, and carbon. In some examples the woody fibre is partially preserved, but it has entirely vanished from others.

In attempting to explain the process of petrifaction in such cases, we may first assume that
strata are very generally permeated by water
charged with minute portions of calcareous, siliceous, and other earths in solution.   In what
manner they become so impregnated will be afterwards considered.   If an organic substance is
exposed in the open air to the action of the sun
and rain, it will in time putrefy, or be dissolved
into its component elements, which consist chiefly
of oxygen, hydrogen, and carbon.   These will
readily be absorbed by the atmosphere or be
washed away by rain, so that all vestiges of the
dead animal or plant disappear.   But if the same
substances be submerged in water, they decompose
more gradually ; and if buried in earth, still more
slowly, as in the familiar example of wooden
piles or other buried timber.   Now, if as fast as
each particle is set free by putrefaction in a fluid
or gaseous state, a particle equally minute of carbonate of lime, flint, or other mineral, is precipitated, we may imagine this inorganic matter
to take the place just before left unoccupied by
the organic molecule.   In this manner a cast of
the interior of certain vessels may first be taken,
and afterwards the walls of the same may decay
and suffer a like transmutation.   Yet when the
whole is lapidified, it may not form one homogeneous mass of stone or metal.   Some of the

original ligneous, osseous, or other organic elements may remain mingled in certain parts, or the lapidifying mineral itself may be so crystallized in different parts as to reflect light differently, and thus the texture of the original body may be faithfully exhibited.

But the student will ask whether, on chemical principles, we have reason to expect that mineral matter will be thrown down precisely in those spots where organic decomposition is in progress? The following curious experiments may serve to illustrate this point. Professor Göppert of Breslau attempted recently to imitate the natural process of petrifaction. For this purpose he steeped a variety of animal and vegetable substances in waters, some holding siliceous, others calcareous, others metallic matter in solution. He found that in the period of a few weeks, or even days, the organic bodies thus immersed were mineralized to a certain extent. Thus, for example, thin vertical slices of deal, taken from the Scotch fir (*Pinus Sylvestris*), were immersed in a moderately strong solution of sulphate of iron. When they had been thoroughly soaked in the liquid for several days, they were dried and exposed to a red-heat until the vegetable matter was burnt up and nothing remained but an oxide of iron, which was found to have taken the form of the deal so exactly that even the dotted vessels peculiar to this family of plants, and resembling those in Fig. 56., were distinctly visible under the microscope.

Another accidental experiment has been recorded by Mr. Pepys in the Geological Transactions. * An earthen pitcher containing several quarts of sulphate of iron had remained undisturbed and unnoticed for about a twelvemonth in the laboratory. At the end of this time when the liquor was examined an oily appearance was observed on the surface, and a yellowish powder, which proved to be sulphur, together with a quantity of small hairs. At the bottom were discovered the bones of several mice in a sediment consisting of small grains of pyrites, others of sulphur, others of crystallized green sulphate of iron, and a black muddy oxide of iron. It was evident that some mice had accidentally been drowned in the fluid, and by the mutual action of the animal matter and the sulphate of iron on each other, the metallic sulphate had been deprived of its oxygen; hence the pyrites and the other compounds were thrown down. Although the mice were not fossilized, or turned into pyrites, the phenomenon shows how mineral waters, charged with sulphate of iron, may be deoxydated on coming in contact with animal matter undergoing putrefaction, so

* Vol. i. p. 399. first series.

that atom after atom of pyrites may be precipitated, and ready, under favourable circumstances, to replace the oxygen, hydrogen, and carbon into which the original body would be resolved.

The late Dr. Turner observes, that when mineral matter is in a " nascent state," that is to say, just liberated from a previous state of chemical combination, it is most ready to unite with other matter, and form a new chemical compound. Probably the particles or atoms just set free are of extreme minuteness, and therefore move more freely, and are more ready to obey any impulse of chemical affinity. Whatever be the cause it clearly follows, as before stated, that where organic matter newly imbedded in sediment is decomposing, there will chemical changes take place most actively.

An analysis was lately made of the water which was flowing off from the rich mud deposited by the Hooghly river in the Delta of the Ganges after the annual inundation. This water was found to be highly charged with carbonic acid gas holding lime in solution. * Now if newly deposited mud is thus proved to be permeated by mineral matter in a state of solution, it is not difficult to perceive that decomposing organic bodies, naturally imbedded in sediment, may as readily become petrified as the substances artificially immersed by Professor Göppert in various fluid mixtures.

* Piddington, Asiat. Research. vol. xviii. p. 226.

It is well known that the water of springs, or
that which is continually percolating the earth's
crust, is rarely free from a slight admixture either
of iron, carbonate of lime, sulphur, flint, potash,
or some other earthy, alkaline, or metallic ingredient. Hot springs in particular are copiously
charged with one or more of these elements; and
it is only in their waters that silex is found in
abundance. In certain cases, therefore, especially
in volcanic regions, we may imagine the flint
of silicified wood and corals to have been supplied by the waters of thermal springs. In other
instances, as in tripoli and chalk-flint, it may
have been derived in great part, if not wholly,
from the decomposition of infusoria, sponges, and
other bodies. But even if this be granted, we
have still to inquire whence a lake or the ocean
can be constantly replenished with the siliceous
matter so abundantly withdrawn from it by the
secretions of these zoophytes.

In regard to carbonate of lime there is no
difficulty, because not only are calcareous springs
very numerous, but even rain-water has the power
of dissolving a minute portion of the calcareous
rocks over which it flows. Hence marine corals
and mollusca may be provided by rivers with the
materials of their shells and solid supports. But
pure silex, even when reduced to the finest powder
and boiled, is insoluble in water. Nevertheless

Dr. Turner has well explained, in an essay on the chemistry of geology*, how the decomposition of felspar may be a source of silex in solution, as widely spread as are the felspathic rocks which form so large a proportion of the volcanic, plutonic, and metamorphic rocks, and are therefore universal, occurring somewhere in the course of every large river.

The siliceous earth, which constitutes more than half the bulk of felspar, is intimately combined with alumine, potash, and some other elements. The alkaline matter of the felspar has a chemical affinity for water, as also for the carbonic acid which is more or less contained in the waters of most springs. The water therefore carries away alkaline matter, and leaves behind a clay consisting of alumine and flint. But this residue of the decomposed mineral, which in its purest state is called porcelain-clay, is found to contain only a small proportion of the silica which existed in the original felspar. The other part therefore must have been dissolved and removed; and this can be accounted for in two ways, first, because silex when combined with an alkali is soluble in water; secondly, because silex in what is technically called its nascent state is also soluble in water. Hence an endless supply of silica is afforded to the waters of the sea.

* Jam. Ed. New Phil. Journ. No. 30. p. 246.

The disintegration of mica also, another mineral which enters largely into the composition of granite and various sandstones, may yield silex which may be dissolved in water, for nearly half of this mineral consists of silica, combined with alumine, potash, and about a tenth part of iron. The oxidation of this iron in the air is the principal cause of the waste of mica.

We have still, however, much to learn before the conversion of fossil bodies into stone is fully understood. Some phenomena seem to imply that the mineralization must proceed with considerable rapidity, for stems of a soft and succulent character, and of a most perishable nature, are preserved in flint; and there are instances of the complete silicification of the young leaves of a palm-tree when just about to shoot forth, and in that state which in the West Indies is called the cabbage of the palm.* It may, however, be questioned whether in such cases there may not have been some antiseptic quality in the water which retarded putrefaction, so that the soft parts of the buried substance may have remained for a long time without disintegration, like the flesh of bodies imbedded in peat.

Mr. Stokes has pointed out examples of petrifactions in which the more perishable, and others

* Stokes, Geol. Trans. vol. v. p. 212. second series.

where the more durable portions of wood are preserved. These variations, he suggests, must doubtless depend on the time when the lapidifying mineral was introduced. Thus in certain silicified stems of palm-trees the cellular tissue, that most destructible part, is in good condition, all signs of the hard woody fibre having disappeared, and the spaces once occupied by it being hollow or filled with agate. Here petrifaction must have commenced soon after the wood was exposed to the action of moisture, and the supply of mineral matter must have failed, or the water have become too much diluted before the woody fibre decayed. But when the latter is alone discoverable, we must then suppose that an interval of time elapsed before the commencement of lapidification, during which the cellular tissue was obliterated. When both structures, namely the cellular and the woody fibre, are preserved, the process must have commenced at an early period, and continued without interruption till it was completed throughout.*

* Stokes, Geol. Trans. vol. v. p. 212. second series.

91

CHAPTER V.

ELEVATION OF STRATA ABOVE THE SEA — HORIZONTAL
AND INCLINED STRATIFICATION.

Why the elevated position of marine strata should be referred to the rising up of the land, not to the going down
of the sea — Upheaval of extensive masses of horizontal
strata — Inclined and vertical stratification — Anticlinal
and synclinal lines — Examples of bent strata in east of
Scotland — Theory of folding by lateral movement — Dip
and strike — Structure of the Jura — Rocks broken by
flexure — Inverted position of disturbed strata — Unconformable stratification — Fractures of strata — Polished
surfaces — Faults — Appearance of repeated alternations
produced by them — Origin of great faults.

LAND has been raised, not the sea lowered. — It has
been already stated that the aqueous rocks containing marine fossils extend over wide continental
tracts, and are seen in mountain chains, rising to
great heights above the level of the sea. Hence
it follows that what is now dry land was once
under water. But if we admit this conclusion we
must imagine either that there has been a general
lowering of the waters of the ocean, or that the
solid rocks, once covered by water, have been raised
up bodily out of the sea, and have thus become
dry land. The earlier geologists, finding themselves reduced to this alternative, embraced the

former opinion, assuming that the ocean was
originally universal, and had gradually sunk down
to its actual level, so that the present islands and
continents were left dry. It seemed to them far
easier to conceive that the water had gone down,
than that solid land had risen upwards into its
present position. It was, however, impossible to
invent any satisfactory hypothesis to explain the
disappearance of so enormous a body of water
throughout the globe, it being necessary to infer
that the ocean had once stood at whatever height
marine. shells were detected. It moreover appeared clear, as the science of Geology advanced,
that certain spaces on the globe had been alternately sea, then land, then estuary, then sea again,
and, lastly, once more habitable land, having remained in each of these states for considerable
periods. In order to account for such phenomena, without admitting any movement of the
land itself, we are required to imagine several
retreats and returns of the ocean ; and even then
our theory applies merely to cases where the
marine strata composing the dry land are horizontal, leaving unexplained those more common
instances where strata are inclined, curved, or
placed on their edges, and evidently not in the
position in which they were first deposited.

Geologists, therefore, were at last compelled to
have recourse to the other alternative, namely, the

doctrine that the solid land has been repeatedly moved upwards or downwards, so as permanently to change its position relatively to the sea. There are several distinct grounds for preferring this conclusion. First, it will account equally for the position of those elevated masses of marine origin in which the stratification remains horizontal, and for those in which the strata are disturbed, broken, inclined, or vertical. Secondly, it is consistent with human experience that land should rise gradually in some places and be depressed in others. Such changes have actually occurred in our own days, and are now in progress, being accompanied in some cases by violent convulsions, while in others they proceed insensibly, or are only ascertainable by the most careful scientific observations. On the other hand, there is no evidence from human experience of a lowering of the sea's level in any region, and the waters of the ocean cannot sink in one place without their level being depressed everywhere throughout the globe.

These preliminary remarks will prepare the reader to understand the great theoretical interest attached to all facts connected with the position of strata, whether horizontal or inclined, curved or vertical.

Now the first and most simple appearance is where strata of marine origin occur above the level

of the sea in horizontal position. Such are the
strata which we meet with in the south of Sicily,
filled with shells of the same species as now live
in the Mediterranean. Some of these rocks rise
to the height of 2000 feet and more above the
sea. Other mountain masses might be mentioned
with horizontal strata of high antiquity which contain fossils wholly dissimilar in form to any now
known to exist, as in the south of Sweden, near
Lake Wener, where the beds of a deposit, called
Transition or Silurian by geologists, occur in as
level a position as if they had recently formed
part of the delta of a great river, and been left
dry on the retiring of the annual floods. Aqueous
rocks of about the same age extend over the
lake-district of North America, exhibiting in like
manner a stratification nearly undisturbed. The
Table Mountain at the Cape of Good Hope is another example of highly elevated and perfectly horizontal strata, no less than 3500 feet in thickness,
and consisting of sandstone of very ancient data.

Instead of imagining that such fossiliferous
rocks were always at their present level, and that
the sea was once high enough to cover them, we
suppose them to have constituted the ancient bed
of the ocean, and that they were gradually uplifted to their present height. This idea, however startling it may at first appear, is quite in
accordance, as above stated, with the analogy of

changes now going on in certain regions of the globe. Thus in parts of Sweden, for example, and the shores and islands of the Gulf of Bothnia, proofs have been obtained that the land is experiencing, and has experienced for centuries, a slow upheaving movement. Playfair argued in favour of this opinion in 1802, and in 1807 Von Buch, after his travels in Scandinavia, announced his conviction that a rising of the land was in progress. Celsius and other Swedish writers had, a century before, declared their belief that a gradual change had for ages been taking place in the relative level of land and sea. They attributed the change to a fall of the waters both of the ocean and the Baltic; but this theory has now been refuted by abundant evidence; for the alteration of relative level has neither been universal nor everywhere uniform in quantity, but has amounted in some regions to several feet in a century, in others to a few inches, while in the southernmost part of Sweden, or the province of Scania, there has been actually a loss instead of a gain of land, buildings having gradually sunk below the level of the sea. *

* In the first three editions of my Principles of Geology, I expressed many doubts as to the validity of the alleged proofs of a gradual rise of land in Sweden; but after visiting that country in 1834 I retracted these objections, and published a detailed statement of the observations which led me to alter my opinion in the Phil. Trans. 1835, Part I. See also Principles, 4th and 5th editions, book ii. chap. 17.

It appears from the observations of Mr. Darwin and others that very extensive regions of the continent of South America have been undergoing slow and gradual upheaval, by which the level plains of Patagonia, covered with recent marine shells, and the Pampas of Buenos Ayres have been formed.* On the other hand the gradual sinking of part of the west coast of Greenland, for the space of more than 600 miles from north to south, during the last four centuries, has been established by the observations of a Danish naturalist, Dr. Pingel. And while these proofs of continental elevation and subsidence by slow and insensible movements have been recently brought to light, the evidence is daily strengthened of continued changes of level effected by violent convulsions in countries where earthquakes are frequent. Here the rocks are rent from time to time, and heaved up or thrown down several feet at once, and disturbed in such a manner, that the original position of strata may, in the course of centuries, be modified to any amount.

It has also been shown by Mr. Darwin, that, in those seas where circular coral islands abound, there is a slow and continued sinking of the submarine mountains on which these masses of coral are based; while in other areas of the South Sea, where coral is found above the sea level, and in

* See his Journal in Voyage of the Beagle.

inland situations, and where there are no circular or barrier reefs, the land is on the rise.*

It would require a volume to explain to the reader the various facts and phenomena which confirm the reality of these movements of land, whether of elevation or depression, whether accompanied by earthquakes or accomplished slowly and without local disturbance. Having treated fully of these subjects in the Principles of Geology, I must assume, in the present work, that such changes are part of the actual course of nature; and when admitted, they will be found to afford a key to the interpretation of a variety of geological appearances, such as the elevation of horizontal or disturbed marine strata, the superposition of freshwater to marine deposits, and many other phenomena, afterwards to be described. It will also appear, in the second part of this volume, how much light the doctrine of a continued subsidence of land may throw on the manner in which a series of strata formed in shallow water may have accumulated to a great thickness. The excavation of valleys also, and other effects of *denudation*, of which I shall presently treat, can alone be understood when we duly appreciate the proofs now on record of the prolonged rising and sinking of land throughout wide areas.

* Proceedings of Geol. Soc. No. 51. p. 552., and his Journal in Voyage of the Beagle, vol. iii. p. 557.

Inclined stratification. — The most unequivocal evidence of a change in the original position of strata is afforded by their standing up perpendicularly on their edges, which is by no means a rare phenomenon, especially in mountainous countries. Thus we find in Scotland, on the southern skirts of the Grampians, beds of puddingstone alternating with thin layers of fine sand, all placed vertically to the horizon. When Saussure first observed certain conglomerates in a similar position in the Swiss Alps, he remarked that the

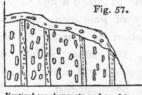

Fig. 57.

Vertical conglomerate and sandstone.

pebbles, being for the most part of an oval shape, had their longer axes parallel to the planes of stratification, (See fig. 57.) From this he inferred that such strata must, at first, have been horizontal, each oval pebble having originally settled at the bottom of the water, with its longer side parallel to the horizon, for the same reason that an egg will not stand on either end if unsupported. Some few, indeed, of the rounded stones in a conglomerate may afford exceptions to the above rule, for the same reason that we see on a shingle beach an occasional oval or flat-sided pebble resting on its end or edge. For some pebbles having been forced along the bottom and against each other, may have settled in this position.

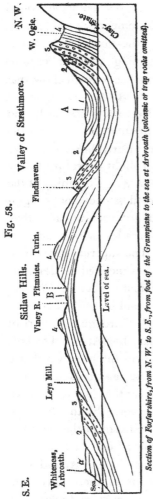

Fig. 58.

Section of Forfarshire, from N. W. to S. E., from foot of the Grampians to the sea at Arbroath (volcanic or trap rocks omitted).

Length of section twenty miles.

Vertical strata, when they can be traced continuously upwards or downwards for some depth, are almost invariably seen to be parts of great curves, which may have a diameter of a few yards, or of several miles. I shall first describe two curves of considerable regularity, which occur in Forfarshire, extending over a country twenty miles in breadth, from the foot of the Grampians to the sea near Arbroath.

The mass of strata here shown may be nearly 2000 feet in thickness, consisting of red and white sandstone, and various coloured shales, the beds being distinguishable into four principal groups, namely, No. 1.

red marl or shale; No. 2. red sandstone, used
for building; No. 3. conglomerate; and No. 4.,
grey paving-stone, and tile-stone, with green
and reddish shale, containing peculiar organic
remains. A glance at the section will show that
each of the formations 2, 3, 4, are repeated
thrice at the surface, twice with a southerly and
once with a northerly inclination or *dip*, and the
beds in No. 1., which are nearly horizontal, are
still brought up twice by a slight curvature to the
surface, once on each side of A.   Beginning at the
north-west extremity, the tile-stones and conglomerates No. 4. and No. 3. are vertical, and they
generally form a ridge parallel to the southern
skirts of the Grampians. The superior strata Nos.
2. and 1. become less and less inclined on descending to the valley of Strathmore, where the strata,
having a concave bend, are said by geologists to lie
in a " trough " or " basin." Through the centre of
this valley runs an imaginary line A, called technically a " synclinal line," where the beds, which are
tilted in opposite directions, may be supposed to
meet. It is most important for the observer to mark
such lines, for he will perceive by the diagram,
that in travelling from the north to the centre of
the basin, he is always passing from older to newer
beds ; whereas after crossing the line A, and pursuing his course in the same southerly direction,
he is continually leaving the newer, and advancing

upon older strata.   All the deposits which he had
before examined begin then to recur in reversed order, until he arrives at the central axis
of the Sidlaw hills, where the strata are seen to
form an arch or *saddle*, having an *anticlinal* line,
B, in the centre.   On passing this line, and continuing towards the S. E., the formations 4, 3,
and 2, are again repeated, in the same relative
order of superposition, but with a northerly dip.
At Whiteness (see diagram) it will be seen that
the inclined strata are covered by a newer deposit, *a*, in horizontal beds.   These are composed
of red conglomerate and sand, and are newer than
any of the groups, 1, 2, 3, 4, before described, and
rest *unconformably* upon strata of the sand-stone
group, No. 2.

An example of curved strata, in which the bends
or convolutions of the rock are sharper and far more
numerous within an equal space, has been well

Fig. 59.

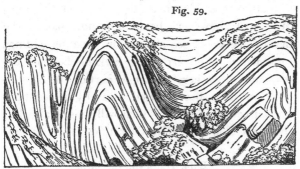

Curved strata of slate near St. Abb's Head, Berwickshire.

F 3

described by Sir James Hall. * It occurs near St. Abb's Head, on the east coast of Scotland, where the rocks consist principally of a bluish slate, having frequently a ripple-marked surface. The undulations of the beds reach from the top to the bottom of cliffs from 200 to 300 feet in height, and there are sixteen distinct bendings in the course of about six miles, the curvatures being alternately concave and convex upwards.

An experiment was made by Sir James Hall, with a view of illustrating the manner in which such strata, assuming them to have been originally horizontal, may have been forced into their present position. A set of layers of clay were placed under a weight, and their opposite ends pressed towards each other with such force as to cause them to approach more nearly together. On the removal of the weight, the layers of clay were found to be curved and folded, so as to bear a miniature resemblance to the strata in the cliffs. We must, however, bear in mind, that in the natural section or sea-cliff we only see the foldings imperfectly, one part being invisible beneath the sea, and the other, or upper portion, being supposed to have been carried away by denudation, or that action of water which will be explained in the next chapter. The dark lines in the accom-

* Edin. Trans. vol. vii. pl. 3.

panying plan (Fig. 60.), represent what is actually
seen of the strata in part of the line of cliff alluded
to ; the fainter lines, that portion which is concealed
beneath the sea level, as also that which is sup-

Fig. 60.

posed to have once existed above the present surface.

We may still more easily illustrate the effects
which a lateral thrust might produce on flexible
strata, by placing several pieces of differently

Fig. 61.

coloured cloths upon a table, and when they are
spread out horizontally, cover them with a book.
Then apply other books to each end, and force

F 4

them towards each other. The folding of the cloths will exactly imitate those of the bent strata. (See Fig. 61.)

To inquire whether the analogous flexures in strata have really been due to similar sideway movements, or other exertions of force, would lead me farther into the regions of speculation and conjecture than might be consistent with the scope of this elementary work. When the volcanic and granitic rocks are described, it will be seen that some of them have, when melted, been injected forcibly into fissures, while others, already in a solid state, have been protruded upwards through the incumbent crust of the earth, by which a great displacement of flexible strata must have been caused. It also appears that cavities are sometimes formed in the interior of the earth, whether by the removal of matter by volcanic action, or by the contraction of argillaceous rocks, or other causes. In this manner pliable beds sinking down, from failure of support, into chasms of less horizontal extent, may have become folded and compressed laterally. Such subsidences have been witnessed on a small scale in the undermined ground immediately over coal-pits, from which large quantities of coal and stone had been extracted.

Between the layers of shale, accompanying coal, we sometimes see the leaves of fossil ferns spread out as regularly as dried plants between sheets of paper in the herbarium of a botanist. These fern leaves, or fronds, must have rested horizontally on soft mud, when first deposited. If, therefore, they and the layers of shale are now inclined, or standing on end, it is obviously the effect of subsequent derangement. The proof becomes, if possible, still more striking when these strata, including vegetable remains, are curved again and again, and even folded into the form of the letter Z, so that the same continuous layer of coal is cut through several times in the same perpendicular shaft. Thus, in the coal-field near Mons, in Bel-

Fig. 62.

Zigzag flexures of coal near Mons.

gium, these zigzag bendings are repeated four or five times, in the manner represented in Fig. 62., the black lines representing seams of coal. *

Dip and Strike. — In the above remarks seve-

* See plan by M. Chevalier, Burat's D'Aubuisson, tom. ii. p. 334.

ral technical terms have been used, such as *dip,*
the *unconformable position* of strata, and the *anticlinal* and *synclinal* lines, which, as well as the
strike of the beds, I shall now explain.   If a stratum or bed of rock, instead of being quite level,
be inclined to one side, it is said to *dip ;* the point
of the compass to which it is inclined is called the
point of dip, and the degree of deviation from a
level or horizontal line is called *the amount of dip,*
or *the angle of dip.*

Fig. 63.

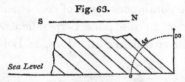

Thus, in the an-
nexed   diagram
(Fig. 63.), a se-
ries of strata are
inclined, and they dip to the north at an angle of
forty-five degrees.   The *strike,* or *line of bearing,*
is the prolongation or extension of the strata in a
direction *at right angles* to the dip ; and hence it is
sometimes called the *direction* of the strata.   Thus,
in the above instance of strata dipping to the
north, their strike must necessarily be east and
west.   We have borrowed the word from the
German geologists, *streichen* signifying to extend,
to have a certain direction.   Dip and strike may
be aptly illustrated by a row of houses running
east and west, the long ridge of the roof representing the strike of the stratum of slates, which dip
on one side to the north, and on the other to the
south.

A stratum which is horizontal, or quite level in all directions, has neither dip nor strike.

It is always important for the geologist, who is endeavouring to comprehend the structure of a country, to learn how the beds dip in every part of the district; but it requires some practice to avoid being occasionally deceived, both as to the point of dip and the amount of it.

If the upper surface of a hard stony stratum be uncovered, whether artificially in a quarry, or by the waves at the foot of a cliff, it is easy to determine towards what point of the compass the slope is steepest, or in what direction water would flow, if poured upon it. This is the true dip. Perfectly horizontal lines in the face of a vertical cliff may be the edges of highly inclined strata, if the observer see them in the line of their strike, their dip being inwards from the face of the cliff. If, however, we come to a break in the cliff, which exhibits a section exactly at right angles to the line of the strike, we are then able to ascertain the true dip. In the annexed drawing (Fig. 64.), we may suppose a headland, one side of which faces to the north, where the beds would appear perfectly horizontal, to a person in the boat; while in the other side facing the west, the true dip would be seen by the person on shore to be at an angle of 40°. If, therefore, our observations are confined to a vertical precipice facing

Fig. 64.

Apparent horizontality of inclined strata.

in one direction, we must endeavour to find a ledge or portion of the plane of one of the beds projecting beyond the others, in order to ascertain the true dip.

It is rarely important to determine the angle of inclination with such minuteness as to require the aid of the instrument called a clinometer. We

Fig. 65.

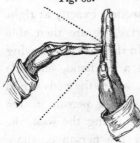

may measure the angle within a few degrees by standing exactly opposite to a cliff where the true dip is exhibited, holding the hands immediately before the eyes, and placing the fingers of one in a perpendicular, and

of the other in a horizontal position, as in Fig. 65. It is thus easy to discover whether the lines of the inclined beds bisect the angle of 90°, formed by the meeting of the hands, so as to give an angle of 45°, or whether it would divide the space into two equal or unequal portions. The upper dotted line may express a stratum dipping to the north; but should the beds dip precisely to the opposite point of the compass as in the lower dotted line, it will be seen that the amount of inclination may still be measured by the hands with equal facility.

It has been already seen, in describing the curved strata on the east coast of Scotland, in Forfarshire and Berwickshire, that a series of concave and convex bendings are occasionally repeated several times. These usually form part of a series of parallel waves of strata, which are prolonged in the same direction throughout a considerable extent of country. Thus, for example, in the Swiss Jura, that lofty chain of mountains has been proved to consist of many parallel ridges, with intervening longitudinal valleys, as in Fig. 66., the ridges being formed by curved fossiliferous strata, of which the nature and dip are occasionally displayed in deep transverse gorges, called "cluses," caused by fractures at right angles to the direction of the chain.*   Now let us suppose these ridges

* See M. Thurmann's work, " Essai sur les Soulèvemens Jurassiques du Porrentruy, Paris, 1832," with whom I examined part of these mountains in 1835.

Fig. 66.

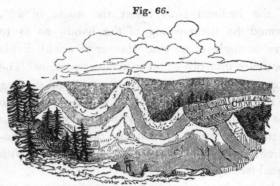

Section illustrating the structure of the Swiss Jura.

and parallel valleys to run north and south, we
should then say that the *strike* of the beds is
north and south, and the *dip* east and west. A
line drawn along the summit of the ridges A, B
would be an anticlinal line, and one following the
bottom of the adjoining valleys a synclinal line.
It will be observed that some of these ridges, A, B,
are unbroken on the summit, whereas one of them,
C, has been fractured along the line of strike,
and a portion of it carried away by denudation,
so that the edges of the beds in the formations
a, b, c come out to the day, or, as the miners say,
crop out, on the sides of a valley. The ground
plan of such a denuded ridge as C may be expressed by the diagram Fig. 67., and the cross
section of the same by Fig. 68. The line D E,
Fig. 67., is the anticlinal line, on each side of
which the dip is in opposite directions, as expressed

Fig. 67.   Fig. 68.

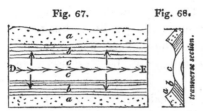

Ground plan of the denuded ridge C, fig. 66.

by the arrows. The emergence of strata at the surface is called by miners their *outcrop* or *basset.*

If, instead of being folded into parallel ridges, the beds form a boss or dome-shaped protuberance, and if we suppose the summit of the dome carried off, the ground plan would exhibit the edges of the strata forming a succession of circles, or ellipses, round a common centre. These circles are the lines of strike, and the dip being always at right angles is inclined in the course of the circuit to every point of the compass, constituting what is termed a qua-qua-versal dip — that is, turning each way.

In the majority of cases, an anticlinal axis forms a ridge, and a synclinal axis a valley, as in A, B, Fig. 58. p. 99.; but there are exceptions to this

Fig. 69.

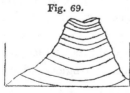

rule, the beds sometimes sloping inwards from either side of a mountain, as in Fig. 69.

On following the anticlinal line of the ridges of the

Jura, before mentioned, A, B, C, Fig. 66., we
often discover longitudinal fissures along the
line where the flexure was greatest.   At the
eastern extremity of the Pyrenees a curious illustration of an analogous phenomenon may be
seen on a small scale (Fig. 70.).   The strata
there laid open, in the sea-cliffs, consist of marl,

Fig. 70.

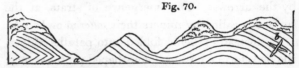

Strata of chert, grit, and marl, near St. Jean de Luz.

grit, and chert, belonging to a formation of
the age of the green-sand of English geologists.
Some of the bendings are so sharp, that fragments of the slaty chert — a hard flinty rock —
taken from the points where they form an angle at
a, might be used for ridge-tiles on the roof of a
house.   Although this chert is now brittle, we
must necessarily suppose that it was flexible when
folded into this shape ; nevertheless it must have
had some solidity, for precisely at the angle of
flexure there are numerous cracks filled with calcedony.   There are also some veins of quartz, *b*,
Fig. 70., traversing the same formation, which
have filled irregular fissures, probably enfiltered at
the same time as the calcedony above mentioned.

Between San Caterina and Castrogiovanni, in
Sicily, bent and undulating gypseous marls occur,

with here and there thin beds of solid gypsum interstratified.   Sometimes these solid layers have been broken into detached fragments, still pre-

Fig. 71.

g. gypsum.   m. marl.

serving their sharp edges, ($g\,g$, Fig. 71.), while the continuity of the more pliable and ductile marls, $m\,m$, has not been inter-rupted.

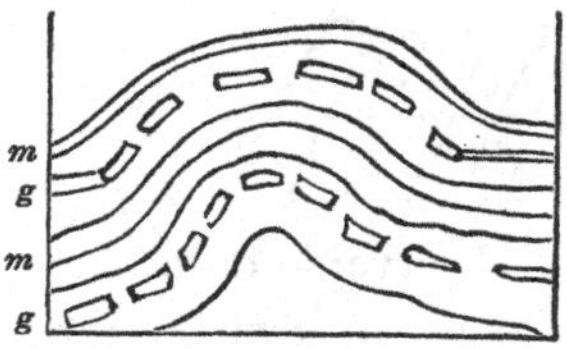

I shall conclude my remarks on bent strata by stating, that, in moun-tainous regions like the Alps, it is often difficult for an experienced geologist to determine cor-rectly the relative age of beds by superposition, so often have the strata been folded back upon themselves, the upper parts of the curve having been removed by denudation.   Thus, if we met with the strata seen in the section Fig. 72., we should naturally suppose that there were twelve

Fig. 72.

distinct beds, or sets of beds, No. 1. being the youngest, and No. 12. the oldest of the series.

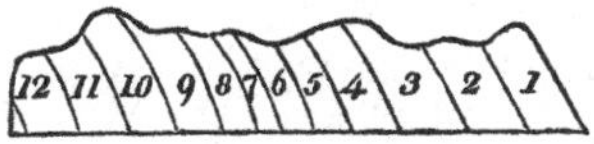

But this section may, perhaps, exhibit merely six beds, which have been folded in the manner seen in Fig. 73., so that each of them are twice re-peated, the position of one half being reversed, and part of No. 1., originally the uppermost, having now become the lowest of the series.

Fig. 73.

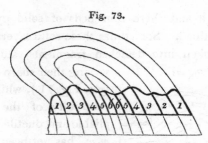

These phenomena are often observable on a magnificent scale in certain regions in Switzerland, where there are precipices from 2000 to 3000 feet in perpendicular height. In the Iselten Alp, in the valley of the Lutschine, between Unterseen and Grindelwald, curves of calcareous shale are seen from 1000 to 1500 feet in height, in which

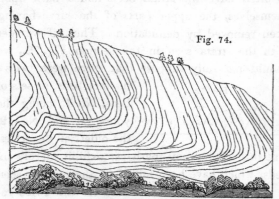

Curved strata of the Iselten Alp.

the beds sometimes plunge down vertically for a depth of 1000 feet and more, before they bend round

again.   There are many flexures not inferior in
dimensions in the Pyrenees, as those near Gavarnie,
at the base of Mont Perdu.

Unconformable stratification. — Strata are said
to be unconformable, when one series is so placed
over another, that the planes of the superior repose on the edges of the inferior.   In this case
it is evident that a period had elapsed between
the production of the two sets of strata, and
that, during this interval, the inferior series
had been tilted and disturbed.   Afterwards the
upper series was thrown down in horizontal strata
upon it.   If these superior beds are also inclined,
it is plain that the lower strata have been twice
displaced; first, when they were themselves brought
into an inclined position, and a second time when
the superior beds were thrown out of the horizontal line.

It often happens that in the interval between
the deposition of two sets of unconformable strata,
the inferior rock has been denuded, and sometimes
drilled by perforating shells.   Thus, for example,
at Autreppe and Gusigny, near Mons, beds of
ancient stone, commonly called transition limestone, highly inclined, and often bent, are covered
with horizontal strata of greenish and whitish marls
of the cretaceous formation, which will be mentioned
in a future chapter.   The lowest and therefore the
oldest bed of the horizontal series is usually the sand

and conglomerate, *a*, in which are rounded frag-

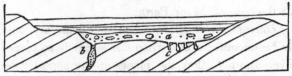

Fig. 75.

Junction of unconformable strata near Mons, in Belgium.

ments of stone, from an inch to two feet in diameter.
These fragments have often adhering shells attached to them, and have been bored by perforating mollusca. The solid surface of the inferior
limestone has also been bored, so as to exhibit
cylindrical and pear-shaped cavities, as at *c*, the
work of saxicavous mollusca; and many rents, as
at *b*, which descend several feet or yards into the
limestone, have been filled with sand and shells,
similar to those in the stratum *a*.

Fractures of the strata. — Numerous rents may
often be seen in rocks which appear to have been
simply broken, the separated parts remaining
in the same places; but we often find a fissure,
several inches or yards wide, intervening between
the disunited portions. These fissures are usually
filled with fine earth and sand, or with angular
fragments of stone, evidently derived from the
fracture of the contiguous rocks.

The face of each wall of the fissure is often
beautifully polished, as if glazed, striated, or
scored with parallel furrows and ridges, such as

would be produced by the continued rubbing together of surfaces of unequal hardness. These polished surfaces are called by miners " slicken-sides." It is supposed that the lines of the striæ indicate the direction in which the rocks were moved. During one of the late minor earthquakes in Chili, the brick walls of a building were rent vertically in several places, and made to vibrate for several minutes during each shock, after which they remained uninjured, and without any open-ing, although the line of each crack was still visible. When all movement had ceased, there were seen on the floor of the house, at the bottom of each rent, small heaps of fine brickdust, evidently pro-duced by trituration.

Faults. — It is not uncommon to find the mass of rock, on one side of a fissure, thrown up above or down below the mass with which it was once in contact on the other side. This mode of displacement is called a shift, slip, or fault. "The miner," says Playfair, describing a fault, " is often perplexed, in his subterraneous journey, by a derangement in the strata, which changes at once all those lines and bearings which had hitherto directed his course. When his mine reaches a certain plane, which is sometimes per-pendicular, as in A B, Fig. 76., sometimes oblique to the horizon (as in C D, ibid.), he finds the beds of rock broken asunder, those on the one side of

Fig. 76.

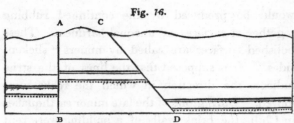

Faults. A B *perpendicular*, C D *oblique to the horizon.*

the plane having changed their place, by sliding in a particular direction along the face of the others. In this motion they have sometimes preserved their parallelism, as in Fig. 76., so that the strata on each side of the faults A B, C D, continue parallel to one another; in other cases, the strata on each side are inclined, as in *a, b, c, d,* (Fig. 77.), though their identity is still to be re-

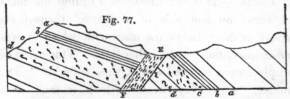

Fig. 77.

E F, *fault or fissure filled with rubbish, on each side of which the shifted strata are not parallel.*

cognized by their possessing the same thickness, and the same internal characters." *

We sometimes see exact counterparts of these slips, on a small scale, in pits of fine loose sand and gravel, many of which have doubtless been caused by the drying and shrinking of argillaceous and

* Playfair, Illust. of Hutt. Theory, § 42.

other beds, slight subsidences having taken place from failure of support. Sometimes, however, even these small slips may have been produced during earthquakes; for land has been moved, and its level, relatively to the sea, considerably altered, within the period when much of the alluvial sand and gravel now covering the surface of continents was deposited.

I have already stated that a geologist must be on his guard, in a region of disturbed strata, against inferring repeated alternations of rocks, when, in fact, the same strata, once continuous, have been bent round so as to recur in the same section, and with the same dip. A similar mistake has often been occasioned by a series of faults.

If, for example, the dark line A H (Fig. 78.) represent the surface of a country on which the strata *a b c* frequently crop out, an observer, who is proceeding from H to A, might at first imagine that at every step he was approaching new strata,

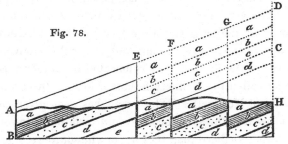

Fig. 78.

Apparent alternations of strata caused by vertical faults.

whereas the repetition of the same beds has
been caused by vertical faults, or downthrows.
Thus, suppose the original mass, A, B, C, D, to
have been a set of uniformly inclined strata, and
that the different masses under E F, F G, and
G D, sank down successively, so as to leave vacant
the spaces marked in the diagram by dotted lines,
and to occupy those marked by the continuous
fainter lines, then let denudation take place
along the line A H, so that the protruding and
triangular masses indicated by the fainter lines
are swept away, — a miner, who has not discovered
the faults, finding the mass a, which we will suppose to be a bed of coal four times repeated,
might hope to find four beds, workable to an indefinite depth, but on arriving at the fault G he
is stopped suddenly in his workings, upon reaching
the strata of sandstone c, or on arriving at the line
of fault F he comes partly upon the shale b, and
partly on the sandstone c, and on reaching E he
is again stopped by a wall composed of the rock d.

The very different levels at which the separated
parts of the same strata are found on the different
sides of the fissure, in some faults, is truly astonish-
ing. One of the most celebrated in England is
that called the " ninety-fathom dike," in the coal-
field of Newcastle. This name has been given to
it, because the same beds are ninety fathoms lower
on the northern than they are on the southern side.

The fissure has been filled by a body of sand, which is now in the state of sandstone, and is called the dike, which is sometimes very narrow, but in other places more than twenty yards wide.* The walls of the fissure are scored by grooves, such as would have been produced if the broken ends of the rock had been rubbed along the plane of the fault. † In the Tynedale and Craven faults, in the north of England, the vertical displacement is still greater, and the horizontal extent of the movement is from twenty to forty miles. Some geologists consider it necessary to imagine that the upward or downward movement in these cases was accomplished at a single stroke, and not by a series of sudden but interrupted movements. This idea appears to have been derived from a notion that the grooved walls have merely been rubbed in one direction. But this is so far from being a constant phenomenon in faults, that it has often been objected to the received theory respecting slickensides, that the striæ are not always parallel, but often curved and irregular. It has, moreover, been remarked, that not only the walls of the fissure or fault, but its earthy contents, sometimes present the same polished and striated faces. Now these facts seem to indicate partial changes

* Conybeare and Phillips, Outlines, &c. p. 376.
† Phillips, Geology, Lardner's Cyclop. p. 41.

G

in the direction of the movement, and some
slidings subsequent to the first filling up of the
fissure. Suppose the mass of rock A, B, C, to
overlie an extensive chasm *d e*, formed at the depth

Fig. 79.

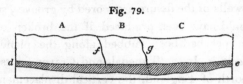

of several miles, whether by the gradual contraction
in bulk of a mass of strata, baked by a moderate
heat, or by the subtraction of matter by volcanic
action, or any other cause.   Now, if this region be
convulsed by earthquakes, the fissures *f g*, and
others at right angles to them, may sever the mass
B from A and from C, so that it may move freely,
and begin to sink into the chasm.   A fracture may
be conceived so clean and perfect as to allow it to
subside at once to the bottom of the subterranean
cavity ; but it is far more probable that the sinking
will be effected at successive periods during different earthquakes, the mass always continuing to
slide in the same direction along the planes of the
fissures *f g*, and the edges of the falling mass being
continually more broken and triturated at each
convulsion.   If, as is not improbable, the circumstances which have caused the failure of support
continue in operation, it may happen that when
the mass B has filled the cavity first formed, its

foundations will again give way under it, so that it will fall again in the same direction. But, if the direction should change, the fact could not be discovered by observing the slickensides, because the last scoring would efface the lines of previous friction. In the present state of our ignorance of the causes of subsidence, an hypothesis which can explain the great amount of displacement in some faults, on sound mechanical principles, by a succession of movements, is far preferable to any theory which assumes each fault to have been accomplished by a single upcast or downthrow of several thousand feet. For we know that there are operations now in progress, at great depths in the interior of the earth, by which both large and small tracts of ground are made to rise above and sink below their former level, some slowly and insensibly, others suddenly and by starts, a few feet or yards at a time; whereas there are no grounds for believing that, during the last 3000 years at least, any regions have been either upheaved or depressed, at a single stroke, to the amount of several hundred, much less several thousand feet.

CHAPTER VI.

DENUDATION, AND THE PRODUCTION OF ALLUVIUM.

Denudation defined — Its amount equal to the entire mass
of stratified deposits in the earth's crust — Horizontal
sandstone denuded in Ross-shire — Levelled surface of
countries in which great faults occur — Connexion of
denudation and alluvial formations — Alluvium, how distinguished from rocks in situ — Ancient alluviums called
diluvium — Origin of these — Erratic blocks and accompanying gravel — Theory of their transportation by ice.

BEFORE we take leave of the aqueous or fossiliferous rocks, we have still to consider the alluvial
formations. *Denudation,* which has been occasionally
spoken of in the preceding chapters, is the removal
of mineral matter by running water, whether by
a river or marine current, and the consequent
laying bare of some inferior rock.   Geologists are,
perhaps, seldom in the habit of reflecting that this
operation is the inseparable accompaniment of the
production of all new strata of mechanical origin.
The transport of sediment and pebbles, to form a
new deposit, necessarily implies that there has
been, somewhere else, a grinding down of rock into
rounded fragments, sand, or mud, equal in
quantity to the new strata.   The gain at one
point has merely been sufficient to balance the
loss at some other.   A ravine, perhaps, has been

excavated, or a valley deepened, or the bed of the sea has, by successive upheaval, been exposed to the power of the waves, so that part of the superior covering of the earth's crust has been stripped off, and thus rocks, previously hidden, have been denuded.

When we see a stone building, we know that somewhere, far or near, a quarry has been opened. The courses of stone in the building may be compared to successive strata, the quarry to a ravine or valley which has suffered denudation. As the strata, like the courses of hewn stone, have been laid one upon another gradually, so the excavation both of the valley and quarry have been gradual. To pursue the comparison still farther, the superficial heaps of mud, sand, and gravel usually called alluvium, may be likened to the rubbish of a quarry which has been rejected as useless by the workmen, or has fallen upon the road between the quarry and the building, so as to lie scattered at random over the ground.

If, then, the entire mass of stratified deposits in the earth's crust is at once the monument and measure of the denudation which has taken place, on how stupendous a scale ought we to find the signs of this removal of transported materials in past ages! Accordingly, there are different classes of phenomena, which attest in a most striking manner the vast spaces left vacant by the erosive power

of water. I may allude first, to those valleys on
both sides of which the same strata are seen following each other in the same order, and having
the same mineral composition and fossil contents.
We may observe for example, several formations,
as Nos. 1, 2, 3, 4, in the accompanying diagram

Fig. 80.

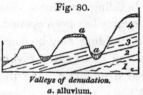

Valleys of denudation.
a. alluvium.

(Fig. 80.); No. 1. conglomerate, No. 2. clay, No. 3.
grit, and No. 4. limestone,
each repeated in a series
of hills separated by valleys varying in depth.
When we examine the subordinate parts of these
four formations, we find, in like manner, distinct
beds in each, corresponding, on the opposite sides
of the valleys, both in composition and order of
position.   No one can doubt that the strata were
originally continuous, and that some cause has
swept away the portions which once connected the
whole series.   A torrent on the side of a mountain
produces similar interruptions, and when we make
artificial cuts in lowering roads, we expose, in like
manner, corresponding beds on either side.   But
in nature, these appearances occur in mountains
several thousand feet high, and separated by intervals of many miles or leagues in extent, of
which a grand exemplification is described by Dr.
MacCulloch, on the north-western coast of Ross-
shire, in Scotland.*

* Western Islands, vol. ii. p. 89. pl. 31. fig. 4.

Fig. 81.

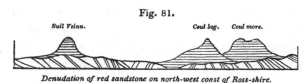

Denudation of red sandstone on north-west coast of Ross-shire.

The fundamental rock of that country is gneiss, in disturbed strata, on which beds of nearly horizontal red sandstone rest unconformably. The latter are often very thin, forming mere flags, with their surface distinctly ripple-marked. They end abruptly on the declivities of many insulated mountains, which rise up at once to the height of about 2000 feet above the gneiss of the surrounding plain or table-land, and to an average elevation of about 3000 feet above the sea, which all their summits generally attain. The base of gneiss varies in height, so that the lower portions of the sandstone occupy different levels, and the thickness of the mass is various, sometimes exceeding 3000 feet. It is impossible to compare these scattered portions without imagining that the whole country has once been covered with a great body of sandstone, and that masses from 1000 to more than 3000 feet in thickness have been removed.

But perhaps the most convincing evidence of denudation on a magnificent scale is derived from the levelled surface of many districts in which large faults occur. I have already shown, in Fig. 78, p. 119, and in Fig. 82, how angular and pro-

truding masses of rock might naturally have been
looked for on the surface immediately above great
faults, although in fact they rarely exist. This
phenomenon may be well studied in those districts
where coal has been extensively worked, for there
the former relation of the beds which have shifted
their position may be determined with great ac-
curacy. Thus in the coal field of Ashby de la
Zouch, in Leicestershire (see Fig. 82.), a fault oc-

Fig. 82.

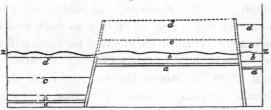

Faults and denuded coal strata, Ashby de la Zouch.

curs, on one side of which the coal beds *a b c d*
rise to the height of 500 feet above the corresponding beds on the other side. But the uplifted
strata do not stand up 500 feet above the
general surface; on the contrary, the outline of
the country, as expressed by the line *z z*, is uniform and unbroken, and the mass indicated by
the dotted outline must have been washed away.*
There are proofs of this kind in some level countries, where dense masses of strata have been

* See Mammat's Geological Facts, &c., p. 90. and plate.

cleared away from areas several hundred square miles in extent.

In the Newcastle coal district, it is ascertained that faults occur in which the upward or downward movement could not have been less than 140 fathoms, which, had they affected equally the configuration of the surface to that amount, would produce mountains with precipitous escarpments near 1000 feet high, or chasms of the like depth; yet is the actual level of the country absolutely uniform, affording no trace whatever of subterraneous disturbance.*

The ground from which these materials have been removed, is usually overspread with heaps of sand and gravel, formed out of the ruins of the very rocks which have disappeared. Thus, in the districts above alluded to, rounded and angular fragments occur of hard sandstone, limestone, and ironstone, with a small quantity of the more destructible shale, and even rounded pieces of coal, the form of these relics pointing to water as the denuding agent.

In geological descriptions we often read of "alluvium" and "diluvium," as opposed to "regular strata," or "fixed rocks," or "rocks in situ." It will be useful, therefore, to explain these terms. At the surface there is commonly a layer of vege-

* Conybeare's Report to Brit. Assoc. 1832. p. 381.

G 5

table mould, derived partly from decayed plants,
and partly caused by the castings of earth-worms,
which are continually sifting the fine from the
coarse soil.*   Immediately beneath this mould
the regular or fundamental stratified or unstratified rocks of the district may appear; but there
usually intervenes, if not an alluvial mass, at least
a quantity of broken and angular fragments of
the subjacent rock, provincially called *rubble*, or
brash, in many parts of England. This last may be
referred partly to the weathering or disintegration of stone on the spot, the effects of air and
water, sun and frost, and chemical decomposition,
and partly to the expanding force of the roots of
trees, which may have grown in small crevices,
at former geological periods, though they may
now be wanting. Sometimes the vibrations and
undulations of earthquakes may have had power,
at some former era, to shatter a surface previously
rent and weathered. Thus in Calabria, subterranean movements have been known to throw up
into the air the slabs of a stone pavement †; and
Mr. Darwin mentions, that in the Island of Qui-
riquina, in Chili, some narrow ridges of hard
primary slate, which is there the fundamental
rock, were as completely shivered by the vibrations

* See Darwin on Formation of Mould, Proceedings of
Geol. Soc. No. 52. p. 574.
† See Princ. of Geol., Index, " Calabria."

of the great earthquake of February 1835, as if
they had been blasted by gunpowder. The effect
was merely superficial, and had caused fresh fractures and displacement of the soil, the slate below
remaining solid and uninjured.*

Alluvium differs from the rubble or brash, just
described, as being composed of sand and gravel,
more or less rolled, in part local, but often in great
part formed of materials transported from a distance. The term is derived from *alluvio*, an inundation, or *alluo*, to wash. The gravel is rarely
consolidated, often unstratified, like heaps of rubbish shot from a cart, but occasionally divided into
wavy and oblique layers, marking successive deposition from water. Such alluvium is strewed
alike over inclined and horizontal strata, and
unstratified rocks ; is most abundant in valleys, but
also occurs in high platforms, and even on lofty
mountains, that of the higher grounds usually
differing from that found at lower levels.

The inferior surface of an alluvial deposit is
often very irregular, conforming to all the inequalities of the subjacent rock. (Fig. 83.) Occasionally a small mass, as at *c*, appears detached,
and as if included in the subjacent formation.
Such isolated portions are usually sections of
winding subterranean hollows filled up with al-

---

* Darwin, p. 370. (for title see note, p. 137.)

Fig. 83.

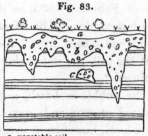

a. vegetable soil.
b. alluvium.
c. mass of same, apparently detached.

luvium. They may have been the courses of springs or subterranean streamlets, which have flowed through and enlarged natural rents; or, when on a small scale, they may be spaces which the roots of large trees have once occupied, gravel and sand having been introduced after their decay.

It is not so easy as may at first appear to draw a clear line of distinction between the *fixed* rocks, or regular strata, (rocks *in situ*, or *in place*), and their alluvial covering of travelled materials. If the bed of a torrent or river be dried up, we call the gravel, sand, and mud, left in their channels, or whatever, during floods, they may have scattered over the neighbouring plains, *alluvium*. The very same materials carried into a lake or sea, where they become sorted by water, and arranged in more distinct layers, are termed regular strata.

In the same manner we may contrast the gravel, sand, and broken shells, strewed along the path of a marine current, with strata formed by the discharge of similar materials, year after year, into a deeper and more tranquil part of the sea.

If any fossils occur, the mass may still be called alluvial, provided the fossils appear to have been drifted to the spot. If any of them, as, for example, freshwater or marine shells, seem to have lived and died where they are entombed, then the deposit, though mainly consisting of drift materials, should not be termed alluvial, but a regular marine or freshwater formation. It is, however, easy to perceive that passages must occur from such alluvial to regular deposits, both in the sea and the estuaries of rivers; and it is often most difficult to distinguish between them, because organic remains have been often obliterated in formations of porous sand, gravel, and loam, which allow rainwater to percolate freely through them.

After what has been said of the connexion of denudation and alluvium, the student will expect to find alluviums of various ages, and at all heights above the sea, formed both before and during the emergence of land, but always most copiously at periods when the level of a country has undergone changes by subterranean movements; for then the course of running water, whether marine or fluviatile, has been most frequently deranged, and the power of the waves of the ocean has been brought to bear with the greatest effect against the land.

Before the doctrine of the rising and sinking of large continental areas, whether insensibly or by a repetition of sudden shocks, was admitted as part of the actual course of nature, all ancient alluviums were classed by some authors under the

common title of " diluvium," and were said not to
be due to existing causes. To establish this pro-
position, it was thought sufficient to demonstrate
that the rivers which may now happen to drain
a given district, could never, in the course of
thousands of ages, have given rise to the valleys of
denudation in which they now flow, and that these
same rivers could never have washed into their
present situations (often the summits of hills, and
high table-lands,) all the gravel and boulders evidently connected with former denuding operations.
It was therefore usual to refer the " diluvium" to
a deluge, or succession of deluges, which rolled with
tremendous violence over the land, after it had
acquired its present configuration, and its present
height above the sea. Not only small gravel, but
large blocks of stone, were supposed to have been
transported from a distance by these devastating
floods or waves, and lodged upon the hill-tops.

But rivers, as we have seen, are not the only
existing causes, nor even the most energetic agents,
by which denudation may be effected. If the
upward movement of land be very slow, the waves
may easily clear away a stratum of yielding materials as fast as they rise, and before they reach
the surface. Thus, a wide uninterrupted ex-
panse of denudation may take place, and masses,
many hundreds of feet or yards in thickness, may
waste away by inches in the course of thousands

of centuries. But if reefs composed of a more refractory stone should at length rise up, the breakers, as they foam over them, may still tear off fragments, and roll them along until the bottom of the sea becomes strewed over with blocks and pebbles. This *alluvium* of marine origin will be uplifted when the reefs are ultimately converted into land, and may then constitute the covering of the summits of hills, or of elevated terraces, or table-lands. At the same time, this gravel may be wanting in all valleys excavated either during the rise of the land, by currents of the sea running between islands, or eaten out or deepened by rivers after the emergence of the land. At the bottom of such more modern valleys a distinct alluvium will be found, containing, perhaps, some pebbles washed out of the older or upland gravel, but principally composed of the ruins of rocks removed during the erosion of the newer valleys.

It must be remembered, that when we introduce such an hypothesis, and take for granted the rise of the land out of the sea, we are merely supposing what we know, from the discovery of marine fossils, to have happened again and again, at former periods.

Erratic blocks. — The great size of the boulders sometimes found associated with ancient alluviums, in places between which and the parent rock deep valleys, and even seas, now intervene, has been thought by some to offer insurmountable objections to any theory which does not introduce causes of great violence to account for their removal. These blocks, called erratic, are some of them a few feet, others several yards, in diameter. They are strewed by myriads over the sandy countries of the north of Germany, and parts of Sweden, Denmark, Finland, and Russia. Some of them at least, must have been carried into their present position since the commencement of a very modern geological period, for they rest, near Stockholm, and elsewhere, on layers of sand and marl containing shells of the species now inhabiting the Baltic.

Although these erratics are far more numerous in northern countries, some are met with as far south as the Swiss Jura, having evidently been carried thither from the Alps, a chain which is now separated from the Jura by one of the broadest and deepest valleys in the world.

Now it is inconceivable how any velocity of water could convey some of these huge masses, over seas and valleys, to the places where they are now found; but there is no real difficulty in supposing them to have been carried by ice, when the lands over which they lie scattered were submerged beneath the sea.

As the reader may perhaps be incredulous respecting the adequacy of the cause here alluded to, I shall enumerate many facts recently brought to light, which incontestably prove how important a part ice plays in the transfer of alluvium from place to place, and especially of that containing large masses of rock. I must confine myself, however, to a brief description of a few examples, as it is not the object of the present work to treat at large of the changes illustrative of geological phenomena, now known to be in progress on the earth.

First, in regard to the distribution of erratics ; they occur, both in the northern and southern hemispheres, between the fortieth parallels of latitude and the poles, but are not met with in the intermediate equatorial and warmer regions.* This fact at once raises a presumption that the greater warmth of parts of Asia, Africa, and America, nearer the line, has proved unfavourable to the transport of such blocks. On the other hand, they abound in the colder regions of North America, from Canada northwards, as well as in northern Europe; and when we travel southwards, and cross the Line in South America, we fall in with them again in Chili and Patagonia, between lat. 41° S. and Cape Horn.† Here, then, we have grounds for suspecting that a cold climate is favourable to the production of erratics.

* See Mr. Darwin on some supposed exceptions to this general rule, Journal of Travels in South America, &c., 1832 to 1836, in Voyage of H. M. S. Beagle, p. 289.

† Darwin, ibid.

Now it is well known, that, annually, in the Baltic, stones are moved by ice; and, very recently, on the shores of the Gulf of Finland, some large fragments were ascertained to have been carried to some distance. In spring, when the fringe of ice which has encircled the coast of the Gulf of Bothnia, and many parts of Sweden, Norway, and Denmark, during winter, breaks up, large stones, with small gravel and ice, which have been firmly frozen into a solid mass on the beach, are floated off to a distance. In Canada similar operations, but on a grander scale, have been noticed by Captain Bayfield. In the river St. Lawrence, the loose ice accumulates on the shoals during winter, at which season the water is low. The separate fragments of ice are readily frozen together in a climate where the temperature is sometimes 30° below zero, and boulders become entangled with them, so that in the spring, when the river rises, on the melting of the snow, the packs are floated off, frequently conveying away the boulders to great distances. A single block of granite, 15 feet long, by 10 feet both in width and height, and which could not contain less than 1500 cubic feet of stone, was in this way moved down the river several hundred yards, during the late survey in 1837. Heavy anchors of ships, lying on the shore have in like manner been closed in and removed. In October, 1836, wooden stakes were driven several feet into the ground, at one point on the banks of the St. Lawrence, at high water mark, and over them were piled many boulders, as large as the united force of six men could roll. The year after, all the boulders had disappeared, and others had arrived, and the stakes had been drawn out and carried away by the ice.

It has also been observed, that ice-islands, detached far to the north, perhaps in Baffin's Bay, are brought by the current, in great numbers, down the coast of Labrador every year, and are often carried through the straits of Belle Isle, between Newfoundland and the continent of America, which, after passing through the straits, sometimes float for several hundred miles to the southwest, up the Gulf of St. Lawrence, between the 40th and 50th degrees of N. latitude. In one of these icebergs, heaps of boulders, gravel, and stones were seen.

A similar agency of ice extends in the southern hemisphere to still lower latitudes. Thus, for example, we learn from Mr. Darwin, that glaciers reaching down to the sea, occur at the head of all the sounds along the western coast of the southern extremity of South America, in latitudes as low as 46°, and still farther, the ice being covered with great fragments of rock. Although these glaciers come down to the sea, the mountains from which they descend have only half the altitude of the Alps, and yet are equidistant from the equator. Portions of this South American ice, charged with

large blocks of granite, were seen in Sir George
Eyre's sound, in the same parallel of latitude as
Paris, floating outwards to the ocean.*

It is therefore natural to suppose that masses
of rock may frequently be carried by icebergs from
the foot of the Andes, in this quarter of South
America, across deep channels, and stranded on
adjacent islands in the Pacific, such as Chiloe,
on which large erratics from the Andes are actually
seen ; and a general elevation of the mainland,
together with the islands, accompanied by the laying dry of the intervening sounds, might present
to a future geologist a problem respecting the
transport of blocks, as enigmatical as any which
are now encountered in Europe.†

Icebergs then, detached from glaciers together
with coast ice, may convey, for hundreds of miles,
pebbles, boulders, sand, and mud, and let these
fall wherever they may chance to melt, on submarine hills and valleys.   These, when the land
emerges from the deep, may constitute some of the
far-transported alluvium which has been ascribed
to diluvial agency.‡

* Darwin, p. 283. (for title, see p. 137.)
† Darwin, ibid. p. 286.
‡ For speculations on the causes of a local and general
change of climate, dependent on fluctuations in physical
geography, and proofs of the wide conversion of sea into
land in Europe, at periods comparatively modern, see Princ.
of Geol. book i.

CHAPTER VII.

VOLCANIC ROCKS.

Trap rocks — Name, whence derived — Their igneous origin at first doubted — Their general appearance and character — Volcanic cones and craters, how formed — Mineral composition and texture of volcanic rocks — Varieties of felspar — Hornblende and augite — Isomorphism — Rocks, how to be studied — Basalt, greenstone, trachyte, porphyry, scoria, amygdaloid, lava, tuff — Alphabetical list, and explanation of names and synonyms, of volcanic rocks — Table of the analyses of minerals most abundant in the volcanic and hypogene rocks.

THE aqueous or fossiliferous rocks having now been described, we have next to examine those which may be called volcanic, in the most extended sense of that term. Suppose *a a* in the annexed

Fig. 84.

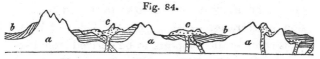

a. Hypogene formations, stratified and unstratified.
b. Aqueous formations.　　*c.* Volcanic rocks.

diagram, to represent the crystalline formations, such as the granitic and metamorphic, *b b* the fossiliferous strata, and *c c* the volcanic rocks. These last are sometimes found, as was explained in the first chapter and Frontispiece, breaking through *a* and *b*, sometimes overlying both, and occasionally alternating with the strata *b b*.   They also are seen, in some instances, to pass insensibly into the unstratified division of *a*, or the Plutonic rocks.

When geologists first began to examine attentively the structure of the northern and western parts of Europe, they were almost entirely ignorant of the phenomena of existing volcanos. They found certain rocks, for the most part without stratification, and of a peculiar mineral composition, to which they gave different names, such as basalt, greenstone, porphyry, and amygdaloid. All these, which were recognized as belonging to one family, were called " trap" by Bergmann (from *trappa*, Swedish, for a flight of steps) — a name since adopted very generally into the nomenclature of the science; for it was observed that many rocks of this class occurred in great tabular masses of unequal extent, so as to form a succession of terraces, or steps, on the sides of hills. This configuration appears to be derived from two causes, first, the abrupt original terminations of sheets of melted matter, which have spread, whether on the land or bottom of the sea, over a level surface. For we know, in the case of lava flowing from a volcano, that a stream, when it has ceased to flow, and grown solid, very commonly ends in a steep slope, as at *a*, Fig. 85.   But, secondly, the step-like

Fig. 85.

Step-like appearance of trap.

appearance arises more frequently from the mode in which horizontal masses of igneous rock, such as *b c*, intercalated between aqueous strata, have, subsequently to their origin, been exposed, at different heights, by denudation. Such an outline, it is true, is not peculiar to trap rocks; great beds of limestone, and other hard kinds of stone, often presenting similar terraces and precipices; but these are usually on a smaller scale, or less numerous, than the volcanic *steps*, or form less decided features in the landscape, as being less distinct in structure and composition from the associated rocks.

Although the characters of trap rocks are greatly diversified, the beginner will easily learn to distinguish them as a class from the aqueous formations. Sometimes they present themselves, as already stated, in tabular masses, which are not divided into strata, sometimes in shapeless lumps and irregular cones, forming small chains of hills. Often they are seen in dikes or wall-like masses, intersecting fossiliferous beds. The rock is occasionally found divided into columns, often decomposing into balls of various sizes, from a few inches to several feet in diameter. The decomposing surface very commonly assumes a coating of a rusty iron colour, from the oxidation of ferruginous matter, so abundant in the traps in which augite or hornblende occur; or, in the felspathic varieties of trap, it acquires a white opaque coating, from the bleaching of the mineral called felspar. On examining any of these volcanic rocks, where they have not suffered disintegration, we rarely fail to detect a crystalline arrangement in one or more of the component minerals. Sometimes the texture of the mass is cellular or porous, or has been porous, and the cells have become filled with carbonate of lime, or other infiltrated mineral, which has thus taken the globular form of the cells.

Most of the volcanic rocks produce a fertile soil by their disintegration. It seems that their component ingredients, silica, alumina, lime, potash, iron, and the rest, are in proportions well fitted for vegetation. As they do not effervesce with acids, a deficiency of calcareous matter might at first have been apprehended; but although carbonate of lime is rare, except in the nodules of amygdaloids, yet it will be seen that lime sometimes enters largely into the composition of augite and hornblende. (See Table, p. 166.)

In regions where the eruption of volcanic matter has taken place in the open air, and where the surface has never since been subjected to great aqueous denudation, cones and craters are strik-

ingly characteristic. Many hundreds of these
cones are seen in central France, in the ancient
provinces of Auvergne, Velay, and Vivarais, where
they observe, for the most part, a linear arrangement, and form chains of hills. Although none of

Fig. 86.

Part of the chain of extinct volcanos called the Monts Dome, Auvergne.
(Scrope.)

the eruptions have happened within the historical
era, the streams of lava may still be traced distinctly descending from many of the craters, and
following the lowest levels of the existing valleys.
The origin of the cone and crater-shaped hill is
well understood, the growth of many having been
watched during volcanic eruptions. A chasm or
fissure first opens in the earth, from which great
volumes of steam and other gases are evolved.
The explosions are so violent as to hurl up into
the air fragments of broken stone, parts of which
are shivered into minute atoms. At the same
time melted stone or *lava* usually ascends through
the chimney or vent by which the gases make
their escape. Although extremely heavy, this
lava is forced up by the expansive power of entangled gaseous fluids, chiefly steam or aqueous

H

vapour, exactly in the same manner as water is made to boil over the edge of a vessel when steam has been generated at the bottom by heat. Large quantities of the lava are also shot up into the air, where it separates into fragments, and acquires a spongy texture by the sudden enlargement of the included gases, and thus forms *scoriæ*, other portions being reduced to an impalpable powder or dust. The showering down of the various ejected materials round the orifice of eruption, gives rise to a conical mound, in which the successive envelopes of sand and scoriæ form layers, dipping on all sides from a central axis. In the mean time a hollow, called a *crater*, has been kept open in the middle of the mound by the continued passage upwards of steam and other gaseous fluids. The lava sometimes flows over the edge of the crater, and thus thickens and strengthens the sides of the cone; but sometimes it breaks it down on one side, and often it flows out from a fissure at the base of the hill. (See Fig. 86.)

I have given a full history and description of the phenomena of recent volcanos in the Principles of Geology, and cannot repeat them here, but shall merely consider the characters of the igneous rocks as they appear to a geologist in the earth's crust. The subject may be treated of in the following order; first, the mineral composition, internal texture, and nomenclature of volcanic rocks;

secondly, the manner and position in which they occur in the earth's crust, and their external forms; and lastly, the connexion between the products of modern volcanos and the rocks usually styled trappean.

Mineral composition and texture. — First, in regard to the composition of volcanic rocks, the varieties most frequently spoken of, are basalt, greenstone, syenitic greenstone, clinkstone, claystone, and trachyte; while those founded chiefly on peculiarities of texture, are porphyry, amygdaloid, lava, tuff, scoriæ, and pumice. It may be stated generally, that all these are mainly composed of two minerals, or families of simple minerals, *felspar* and *hornblende*, some almost entirely of hornblende, others of felspar.

These two minerals may be regarded as two groups, rather than species. Felspar, for example, may be, first, common felspar, that is to say, potash-felspar, in which the alkali is potash (*see* Table, p. 166.); or, secondly, albite, that is to say, soda-felspar, where the alkali is soda instead of potash; or, thirdly, Labrador-felspar (Labradorite), which differs not only in its iridescent hues, but also in its angle of fracture or cleavage, and its composition. We also read much of two other kinds, called glassy felspar and compact felspar, which, however, cannot rank as varieties of equal importance, for both the albitic and common felspar appear sometimes in transparent or *glassy* crystals; and as to compact felspar, it is probably a compound of a less definite nature, sometimes containing, according to Dr. MacCulloch, both soda and potash.

The other group, or *hornblende*, consists principally of two varieties; first, hornblende, and, secondly, augite, which were once regarded as very distinct, although now some eminent mineralogists are in doubt whether they are not one and the same mineral, differing only as one crystalline form of native sulphur differs from another.

The history of the changes of opinion on this point is curious and instructive. Werner first distinguished augite from hornblende; and his proposal to separate them obtained afterwards the sanction of Haüy, Mohs, and other celebrated mineralogists. It was agreed that the form of the crystals of the two species were different, and their structure, as shown by *cleavage*, that is to say, by breaking or cleaving the mineral with a chisel, or a blow of the hammer, in the direction in which it yields most readily. It was also found by analysis that augite usually contained more lime, less alumina, and no fluoric acid; which last, though not always found in hornblende, often enters into its composition, in minute quantity. In addition to these characters, it was remarked as a geological fact, that augite and hornblende are very rarely associated together in the same rock; and that

when this happened, as in some lavas of modern
date, the hornblende occurs in the mass of the
rock, where crystallization may have taken place
more slowly, while the augite merely lines cavities
where the crystals may have been produced rapidly
It was also remarked, that in the crystalline slags
of furnaces, augitic forms were frequent, the hornblendic entirely absent; hence it was conjectured
that hornblende might be the result of slow, and
augite of rapid cooling. This view was confirmed by the fact, that Mitscherlich and Berthier
were able to make augite artificially, but could
never succeed in forming hornblende. Lastly,
Gustavus Rose fused a mass of hornblende in a
porcelain furnace, and found that it did not, on
cooling, assume its previous shape, but invariably
took that of augite. The same mineralogist observed certain crystals in rocks from Siberia which
presented a hornblende *cleavage*, while they had
the external form of augite.

If, from these data, it is inferred that the same
substance may assume the crystalline forms of
hornblende or augite indifferently, according to
the more or less rapid cooling of the melted mass,
it is nevertheless certain that the variety commonly called augite, and recognized by a peculiar
crystalline form, has usually more lime in it, and
less alumina, than that called hornblende, although
the quantities of these elements do not seem to be

always the same. Unquestionably the facts and experiments above mentioned show the very near affinity of hornblende and augite; but even the convertibility of one into the other by melting and recrystallizing, does not perhaps demonstrate their absolute identity. For there is often some portion of the materials in a crystal which are not in perfect chemical combination with the rest. Carbonate of lime, for example, sometimes carries with it a considerable quantity of silex into its own form of crystal, the silex being mechanically mixed as sand, and yet not preventing the carbonate of lime from assuming the form proper to it. This is an extreme case, but in many others some one or more of the ingredients in a crystal may be excluded from perfect chemical union; and after fusion, when the mass recrystallizes, the same elements may combine perfectly or in new proportions, and thus a new mineral may be produced. Or some one of the gaseous elements of the atmosphere, the oxygen for example, may, when the melted matter reconsolidates, combine with some one of the component elements.

The different quantity of the impurities or refuse above alluded to, which may occur in all but the most transparent and perfect crystals, may partly explain the discordant results at which experienced chemists have arrived in their analysis of the same mineral. For the reader will find that a mineral

determined to be the same by its physical characters, crystalline form, and optical properties,
has often been declared by skilful analysers to
be composed of distinct elements. (See the Table
at p. 166.) This disagreement seemed at first sub-
versive of the doctrine, that there is a fixed and
constant relation between the crystalline form and
structure of a mineral, and its chemical composition. The apparent anomaly, however, which
threatened to throw the whole science of mineralogy into confusion, was in a great degree recon-
ciled to fixed principles by the discoveries of Professor Mitscherlich at Berlin, who ascertained that
the composition of the minerals which had appeared so variable, was governed by a general
law, to which he gave the name of *isomorphism*
(from σος, *isos*, equal, and μορφη, *morphe*, form).
According to this law, the ingredients of a given
species of mineral are not absolutely fixed as to
their kind and quality; but one ingredient may be
replaced by an equivalent portion of some analogous ingredient. Thus, in augite, the lime may
be in part replaced by portions of protoxide of
iron, or of manganese, while the form of the crystal, and the angle of its cleavage planes, remain
the same. These vicarious substitutions, however,
of particular elements cannot exceed certain defined limits.

Having been led into this digression on the recent progress of mineralogy, I may here observe that the geological student must endeavour as soon as possible to familiarize himself with the characters of five at least of the most abundant simple minerals of which rocks are composed. These are, felspar, quartz, mica, hornblende, and carbonate of lime   This knowledge cannot be acquired from books, but requires personal inspection, and the aid of a teacher.   It is well to accustom the eye to know the appearance of rocks under the lens.   To learn to distinguish felspar from quartz is the most important step to be first aimed at; when these occur in a granular and uncrystallized state, the young geologist must not be discouraged if, after considerable practice, he often fails to distinguish them by the eye alone.   If the felspar is in crystals, it is easily recognized by its cleavage; but when in grains the blow-pipe must be used, for the edges of the grains can be rounded in the flame, whereas those of *quartz* are infusible.   If the geologist is desirous of distinguishing the three varieties of felspar above enumerated, or hornblende from augite, it will often be necessary to use the reflecting goniometer as a test of the angle of cleavage, and shape of the crystal.   The use of this instrument will not be found difficult.

The external characters and composition of the felspars are extremely different from those of augite or hornblende; so that the volcanic rocks in

which either of these minerals decidedly predominate, are easily recognized.   But there are
mixtures of the two elements in every possible
proportion, the mass being sometimes exclusively
composed of felspar, at other times solely of
augite, or, again, of both in equal quantities. Occasionally, the two extremes, and all the intermediate gradations, may be detected in one continuous
mass.   Nevertheless there are certain varieties or
compounds which prevail so largely in nature, and
preserve so much uniformity of aspect and composition, that it is useful in geology to regard them
as distinct rocks, and to assign names to them,
such as basalt, greenstone, trachyte, and others,
already mentioned.

   Basalt. — As an example of rocks in which augite greatly prevails, basalt may first be mentioned.
Although we are more familiar with this term
than with that of any other kind of trap, it is difficult to define it, the name having been used
so vaguely.   It has been very generally applied to
any trap rock of a black, bluish, or leaden-grey
colour, having a uniform and compact texture.
Most strictly, it consists of an intimate mixture of
augite, felspar, and iron, to which a mineral of an
olive green colour, called olivine, is often superadded, in distinct grains or nodular masses.   The
iron is usually magnetic, and is often accompanied
by another metal, titanium.   Augite is the predo-

minant mineral, the felspar being in much smaller proportions.   There is no doubt that many of the fine-grained and dark-coloured trap rocks, called basalt, contain hornblende in the place of augite; but this will be deemed of small importance after the remarks above made.   Other minerals are occasionally found in basalt; and this rock may pass insensibly into almost any variety of trap, especially into greenstone, clinkstone, and wacké, which will be presently described.

Greenstone, or *Dolerite*, is usually defined as a granular rock, the constituent parts of which are hornblende and imperfectly crystallized felspar; the felspar being more abundant than in basalt, and the grains or crystals of the two minerals more distinct from each other.   This name may also be applied when augite is substituted for hornblende (the dolerite of some authors), or when albite replaces common felspar, forming the rock sometimes called Andesite.

Syenitic greenstone. —The highly crystalline compounds of the same two minerals, felspar and hornblende, having a granitiform texture, and with occasionally some quartz accompanying, may be called Syenitic greenstone, a rock which frequently passes into ordinary trap, and as frequently into granite.

Trachyte. —A porphyritic rock of a whitish or greyish colour, composed principally of glassy felspar, with crystals of the same, generally with some hornblende and some titaniferous iron. In composition it is extremely different from basalt, this being a felspathic, as the other is an augitic, rock. It has a peculiar rough feel, whence the name τραχυς, *trachus*, rough. Some varieties of trachyte contain crystals of quartz.

Porphyry is merely a certain form of rock, very characteristic of the volcanic formations. When distinct crystals of one or more minerals are scattered through an earthy or compact base, the rock is termed a porphyry. (See Fig. 87.) Thus trachyte is porphyritic; for in it, as in many mo-

Fig. 87.

Porphyry.
White crystals of felspar in a dark base of hornblende and felspar.

dern lavas, there are crystals of felspar; but in some porphyries the crystals are of augite, olivine, or other minerals. If the base be greenstone, basalt, or pitchstone, the rock may be denominated greenstone-porphyry, pitchstone-porphyry, and so forth.

Amygdaloid. — This is also another form of igneous rock, admitting of every variety of composition. It comprehends any rock in which round or almond-shaped nodules of some mineral, such as agate, calcedony, calcareous spar, or zeolite, are

scattered through a base of wacké, basalt, greenstone, or other kind of trap. It derives its name
from the Greek word *amygdala*, an almond. The
origin of this structure cannot be doubted, for we
may trace the process of its formation in modern
lavas. Small pores or cells are caused by bubbles
of steam and gas confined in the melted matter.
After or during consolidation these empty spaces
are gradually filled up by matter separating from
the mass, or infiltered by water permeating the
rock. As these bubbles have been sometimes
lengthened by the flow of the lava before it finally
cooled, the contents of such cavities have the form
of almonds. In some of the amygdaloidal traps of
Scotland, where the nodules have decomposed, the
empty cells are seen to have a glazed or vitreous
coating, and in this respect exactly resemble scoriaceous lavas, or the slags of furnaces.

The annexed figure represents a fragment of

Fig. 88.

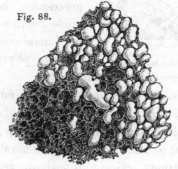

Scoriaceous lava in part converted into an amygdaloid. — Montagne
de la Veille, Department of Puy de Dome, France.

stone taken from the upper part of a sheet of basaltic lava in Auvergne. One half is scoriaceous, the pores being perfectly empty, the other part is amygdaloidal, the pores or cells being mostly filled up with carbonate of lime, forming white kernels.

Scoriæ and *Pumice* may next be mentioned as porous rocks, produced by the action of gases on materials melted by volcanic heat. *Scoriæ* are usually of a reddish brown and black colour, and are the cinders and slags of basaltic or augitic lavas. *Pumice* is a light, spongy, fibrous substance, produced by the action of gases on trachytic and other lavas; the relation, however, of its origin to the composition of lava is not yet well understood. Von Buch says it does not occur where only Labrador-felspar is present.

Lava. — This term has a somewhat vague signification, having been applied to all melted matter observed to flow in streams from volcanic vents. When this matter consolidates in the open air, the upper part is usually scoriaceous, and the mass becomes more and more stony as we descend, or in proportion as it has consolidated more slowly and under greater pressure. At the bottom, however, of a stream of lava, a small portion of scoriaceous rock very frequently occurs, formed by the first thin sheet of liquid matter, which often precedes the main current, or by contact with water in or upon the damp soil.

The more compact lavas are often porphyritic, but even the scoriaceous part sometimes contains imperfect crystals, which have been derived from some older rocks, in which the crystals pre-existed, but were not melted, as being more infusible in their nature.

Although melted matter rising in a crater, and even that which enters rents on the side of a crater, is called lava, yet this term belongs more properly to that which has flowed either in the open air or on the bed of a lake or sea. If the same fluid has not reached the surface, but has been merely injected into fissures below ground, it is called trap.

There is every variety of composition in lavas; some are trachytic, as in the Peak of Teneriffe; a great number are basaltic, as in Vesuvius and Auvergne; others are Andesitic, as those of Chili; some of the most modern in Vesuvius consist of green augite, and many of those of Etna of augite and Labrador-felspar.*

Trap tuff, volcanic tuff. — Small angular fragments of the scoriæ and pumice, above mentioned, and the dust of the same, produced by volcanic explosions, form the tuffs which abound in all regions of active volcanos, where showers of these materials, together with small pieces of other

* G. Rose, Ann. des Mines, tom. 8. p. 32.

rocks ejected from the crater, fall down upon the land or into the sea. Here they often become mingled with shells, and are stratified. Such tuffs are sometimes bound together by a calcareous cement, and form a stone susceptible of a beautiful polish. But even when little or no lime is present, there is a great tendency in the materials of ordinary tuffs to cohere together.

Besides the peculiarity of their composition, some tuffs, or *volcanic grits*, as they have been termed, differ from ordinary sandstones by the angularity of their grains. When the fragments are coarse, the rock is styled a volcanic *breccia*. *Tufaceous conglomerates* result from the intermixture of rolled fragments or pebbles of volcanic and other rocks with tuff.

According to Mr. Scrope, the Italian geologists confine the term *tuff*, or tufa, to felspathose mixtures, and those composed principally of pumice, using the term *peperino* for the basaltic tuffs.*

We meet occasionally with extremely compact beds of volcanic materials, interstratified with fossiliferous rocks, much resembling the trap which may be found in a dike. These may sometimes be tuffs, notwithstanding their density or compactness. The chocolate-coloured mud, which

* Geol. Trans. vol. ii. p. 211. Second Series.

was poured for weeks out of the crater of Graham's Island, in the Mediterranean, in 1831, must, when unmixed with other materials, have constituted a stone heavier than granite. Each cubic inch of the impalpable powder which has fallen for days through the atmosphere during some modern eruptions, has been found to weigh, without being compressed, as much as ordinary trap rocks, which are often identical in mineral composition.

The fusibility of the igneous rocks generally exceeds that of other rocks, for there is much alkaline matter and lime in their composition, which serves as a flux to the large quantity of silica, which would be otherwise so refractory an ingredient.

It is remarkable, that notwithstanding the abundance of this silica, quartz is wanting in the volcanic rocks, or is present only as an occasional mineral, like mica. The elements of mica, as of quartz, occur in lava and trap, but the circumstances under which these rocks are formed, are evidently unfavourable to the development of mica and quartz, minerals so characteristic of the hypogene formations.

It would be tedious to enumerate all the varieties of trap and lava which have been regarded by different observers as sufficiently abundant to deserve distinct names, especially as each investigator is too apt to exaggerate the importance of local varieties which happen to prevail in districts best known to him. It will be useful, however, to subjoin here, in the form of a glossary, an alphabetical list of the names and synonyms most commonly in use, with brief explanations, to which I have added a table of the analysis of the simple minerals most abundant in the volcanic and hypogene rocks.

Explanation of the names, synonyms, and mineral composition of the more abundant volcanic rocks.

AMPHIBOLITE. *See* Hornblende rock, amphibole being Haüy's name for hornblende.

AMYGDALOID. A particular form of volcanic rock; *see* p. 155.

AUGITE ROCK. A kind of basalt or greenstone, composed wholly or principally of granular augite. (*Leonhard's Mineralreichs,* 2d edition, p. 85.)

AUGITIC-PORPHYRY. Crystals of Labrador-felspar and of augite, in a green or dark grey base. (*Rose, Ann. des Mines,* tom. 8. p. 22. 1835.)

BASALT. Chiefly augite — an intimate mixture of augite and felspar with magnetic iron, olivine, &c. *See* p. 153. The yellowish green mineral called olivine, can easily be distinguished from yellowish felspar by its infusibility, and having no cleavage. The edges turn brown in the flame of the blow-pipe.

CLAYSTONE and CLAYSTONE-PORPHYRY. An earthy and compact stone, usually of a purplish colour, like an indurated clay;

passes into hornstone; generally contains scattered crystals of felspar and sometimes of quartz.

CLINKSTONE. *Syn.* Phonolite, fissile Petrosilex; a greenish or greyish rock, having a tendency to divide into slabs and columns; hard, with clean fracture, ringing under the hammer; principally composed of compact felspar, and, according to Gmelin, of felspar and mesotype. (*Leonhard, Mineralreichs,* p. 102.) A rock much resembling clinkstone, and called by some Petrosilex, contains a considerable percentage of quartz and felspar. As both trachyte and basalt pass into clinkstone, the rock so called must be very various in composition.

COMPACT FELSPAR, which has also been called Petrosilex; the rock so called includes the hornstone of some mineralogists, is allied to clinkstone, but is harder, more compact, and translucent. It is a varying rock, of which the chemical composition is not well defined, and is perhaps the same as that of clay. (*MacCulloch's Classification of Rocks,* p. 481.) Dr. MacCulloch says, that it contains both potash and soda.

CORNEAN. A variety of claystone allied to hornstone. A fine homogeneous paste, supposed to consist of an aggregate of felspar, quartz, and hornblende, with occasionally epidote, and perhaps chlorite; it passes into compact felspar and hornstone. (*De la Beche, Geol. Trans.* second series, vol. 2. p. 3.)

DIALLAGE ROCK. *Syn.* Euphotide, Gabbro, and some Ophiolites. Compounded of felspar, and diallage, sometimes with the addition of serpentine, or mica, or quartz. (*MacCulloch, ibid.* p. 648.)

DIORITE. A kind of greenstone, which see. Components, felspar and hornblende in grains. According to *Rose, Ann. des Mines,* tom. 8. p. 4., *diorite* consists of albite and hornblende.

DIORITIC-PORPHYRY. A porphyritic greenstone, composed of crystals of albite and hornblende, in a greenish or blackish base. (*Rose, ibid.* p. 10.)

DOLERITE. Formerly defined as a synonym of greenstone, which see. But according to Rose (*ibid.* p. 32.), its composition is black augite and Labrador-felspar; according to Leonhard

(*Mineralreichs*, &c. p. 77.), augite, Labrador-felspar, and magnetic iron.

DOMITE. An earthy condition of *trachyte*, found in the Puy de Dome, in Auvergne.

EUPHOTIDE. A mixture of grains of Labrador-felspar and diallage. (*Rose, ibid.* p. 19.) According to some, this rock is defined to be a mixture of augite or hornblende, and Saussurite, a mineral allied to jade. (*Allan's Mineralogy*, p. 158.) *See* Diallage rock.

FELSPAR-PORPHYRY. *Syn.* Hornstone-porphyry; a base of felspar, with crystals of felspar, and crystals and grains of quartz. *See* also Hornstone.

GABBRO, *see* Diallage-rock.

GREENSTONE; *Syn.* Dolerite and diorite; components, hornblende and felspar, or augite and felspar in grains. See above, p. 154.

GREYSTONE. (Graustein of Werner.) Lead grey and greenish rock, composed of felspar and augite, the felspar being more than seventy-five per cent. (*Scrope, Journ. of Sci.* No. 42. p. 221.) Greystone lavas are intermediate in composition between basaltic and trachytic lavas.

HORNBLENDE ROCK. A greenstone, composed wholly or principally of granular hornblende, or augite. (*Leonhard, Mineralreichs*, &c., p. 85.)

HORNSTONE, HORNSTONE-PORPHYRY. A kind of felspar porphyry, (*Leonhard, ibid.*) with a base of hornstone, a mineral approaching near to flint, differing from compact felspar in being infusible.

HYPERSTHENE ROCK, a mixture of grains of Labrador-felspar and hypersthene, (*Rose, Ann. des Mines*, tom. 8. p. 13.), having the structure of syenite or granite; abundant among the traps of Sky. In a geological view, it has been called a greenstone, in which hypersthene takes the place of hornblende.

MELAPHYRE. A variety of black porphyry, the base being black augite with crystals of felspar; from μελας, *melas*, black.

OBSIDIAN. Vitreous lava like melted glass, nearly allied to pitchstone.

OPHIOLITE, sometimes same as Diallage rock (*Leonhard*, p. 77.); sometimes a kind of serpentine.

OPHITE. A green porphyritic rock, composed chiefly of hornblende, with crystals of that mineral in a base of the same, mixed with some felspar. It passes into serpentine by a mixture of talc. (*Burat's D'Aubuisson*, tom. 2. p. 63.)

PEARLSTONE. A volcanic rock having the lustre of mother of pearl; usually having a nodular structure; intimately related to obsidian, but less glassy.

PEPERINO. A form of volcanic tuff, composed of basaltic scoriæ See p. 159.

PETROSILEX. *See* Clinkstone and Compact Felspar.

PHONOLITE. *Syn.* of Clinkstone, which see.

PITCHSTONE; vitreous lava, less glassy than obsidian; a blackish green rock resembling glass, having a resinous lustre and appearance of pitch; composition various, usually felspar and augite; passes into basalt; occurs in veins, and in Arran forms a dike thirty feet wide, cutting through sandstone; forms the outer walls of some basaltic dikes.

PORPHYRY. Any rock in which detached crystals of felspar, or of one or more minerals, are diffused through a base. *See* p. 155.

POZZOLANA. A kind of tuff. *See* p. 75.

PUMICE. A light, spongy, fibrous form of trachyte. *See* p. 157.

PYROXENIC-PORPHYRY, same as augitic-porphyry, pyroxene being Haüy's name for augite.

SCORIÆ. *Syn.* volcanic cinders; reddish brown or black porous form of lava. *See* p. 157.

SERPENTINE. A greenish rock, in which there is much magnesia; usually contains diallage, which is nearly allied to the simple mineral called serpentine. Occurs sometimes, though rarely, in dikes, altering the contiguous strata; is indifferently a member of the trappean or hypogene series.

SYENITIC-GREENSTONE; composition, crystals or grains of felspar and hornblende. *See* p. 154.

TEPHRINE, synonymous with lava.

TOADSTONE.   A local name in Derbyshire for a kind of wacké, which see.

TRACHYTE, chiefly composed of glassy felspar, with crystals of glassy felspar.   *See* p. 154.

TRAP TUFF.   *See* p. 158.

TRASS.   A kind of tuff or mud poured out by lake-craters during eruptions; common in the Eifel, in Germany.

TUFACEOUS CONGLOMERATE.   *See* p. 159.

TUFF.   *Syn.* Trap-tuff, volcanic tuff.   *See* p. 158.

VITREOUS LAVA.   *See* Pitchstone and Obsidian.

VOLCANIC TUFF.   *See* p. 158.

WACKÉ.   A soft and earthy variety of trap, having an argillaceous aspect.   It resembles indurated clay, and when scratched, exhibits a shining streak.

WHINSTONE.   A Scotch provincial term for greenstone and other hard trap rocks.

ANALYSIS OF MINERALS MOST ABUNDANT IN THE VOLCANIC AND HYPOGENE ROCKS.

	Silica.	Alumina.	Magnesia.	Lime.	Potash.	Soda.	Iron Oxide.	Manganese.	Remainder.
Actinolite (Bergman) -	64·	-	22·	-	-	-	3·	-	-
Albite (Rose) -	68·84	20·53	-	a trace	-	9·12	-	-	-
—— (mean of 4 analyses)	69·45	19·44	0·13	0·22	-	9·95	a trace	a trace	-
Augite (Rose) -	53·36	-	4·99	22·19	-	-	17·38	0·09	-
—— (mean of 4 analyses)	53·57	1·	11·26	20·9	-	-	10·75	0·67	-
Carbonate of Lime (Biot.) -	-	-	-	56·33	-	-	-	-	43·05 C.
Chiastolite (Landgrabe) -	68·49	30·17	4·12	-	-	-	2·7	-	0·27 W.
Chlorite (Vauquelin) -	26·	18·5	8.	-	-	2·	43·	-	-
—— (mean of 3 analyses)	27·43	17·9	14·56	0·50	1·56	-	30·63	-	6·92 W.
Diallage (Klaproth) -	60·	-	27·5	-	-	-	10·5	-	-
—— (mean of 3 analyses)	43·33	2·2	26·41	5·58	-	-	11·53	-	8·54 W.
Epidote (Vauquelin) -	37·	21·	-	15·	-	-	24·	1·5	-
Felspar, common, (Vauq.) -	62·83	17·02	-	3·	13·	-	1·	-	-
—— (Rose) -	66·75	17·5	-	1·25	12·	-	0·75	-	-
—— (mean of 7 analyses)	64·04	18·94	-	0·76	13·66	-	0·74	-	-
Garnet (Klaproth) -	35·75	27·25	-	-	-	-	36·	0·25	-
—— (Phillips) -	43·	16·	-	20·	-	-	16·	-	-

	Silica.	Alumina.	Magnesia.	Lime.	Potash.	Soda.	Iron Oxide.	Manganese.	Remainder.
Hornblende (Klap.) - -	42·	12·	2·25	11·	a trace	-	30·	0·25	1·5 F.
—— (Bonsdorff.) -	45·69	12·18	18·79	13·85	-	-	7·32	0·22	1· W.
Hypersthene (Klaproth) -	54·25	2·25	14·	1·5	-	-	24·5	a trace	0·5 W.
Labrador-felspar, (Klap.)	55·75	26·5	-	11·	-	4·	1·25	-	
Leucite (Klap.) - -	53·75	24·62	-	-	21·35	-	-	-	
Mesotype (Gehlen) -	54·46	19·70	-	1·61	-	15·09	-	-	9·83 W.
Mica (Klaproth) -	42·5	11·5	9·	-	10·	-	22·	-	
—— (Vauquelin) -	50·	35·	-	1·33	-	-	7·	2·	
—— (mean of 3 analyses)	45·83	22·58	-	-	11·08	-	14·	1·45	
Olivine (Klaproth) -	50·	-	38·5	-	-	-	12·	-	
Schorl or Tourmaline (Gmelin)	35·48	34·75	4·68	-	0·48	1·75	17·44	1·89	4·02 B.
—— (mean of 6 analyses)	36·03	35·82	4·44	0·28	0·71	1·96	13·71	1·62	12·45 W.
Serpentine (Hisinger) -	43·07	0·25	40·37	0·5	-	-	1·17	-	12·77 W.
—— (mean of 5 analyses)	37·29	4·97	36·8	2·89	-	-	3·14	-	5· W.
Steatite (Vauquelin) -	64·	-	22·	-	-	-	3·	-	
—— (mean of 3 anal. by Klap.)	48·3	6·18	26·65	-	-	-	2·	-	9·5 W.
Talc (Klaproth) -	61·75	-	30·5	-	2·75	-	2·5	-	

In the last column of the above Table, the letters B. C. F. W. represent Boracic acid, Carbonic acid, Fluoric acid, and Water.

CHAPTER VIII.

HAVING in the last chapter spoken of the composition and mineral characters of volcanic rocks,
I shall next describe the manner and position in
which they occur in the earth's crust, and their
external forms. Now the leading varieties, such
as basalt, greenstone, trachyte, porphyry, and the
rest are found sometimes in dikes penetrating stratified and unstratified formations, sometimes in
shapeless masses protruding through or overlying
them, or in horizontal sheets intercalated between
strata.

Volcanic dikes. — Fissures have already been
spoken of as occurring in all kinds of rocks, some
a few feet, others many yards in width, and often

filled up with earth or angular pieces of stone, or
with sand and pebbles. Instead of such materials,
suppose a quantity of melted stone to be driven
or injected into an open rent, and there consolidated, we have then a tabular mass resembling a
wall, and called a trap dike. It is not uncommon
to find such dikes passing through strata of soft
materials, such as tuff or shale, which, being more
perishable than the trap, are often washed away
by the sea, rivers, or rain, in which case the dike
stands prominently out in the face of precipices,
or on the level surface of a country. (See the
annexed figure.) *

Fig. 89.

Dike in inland valley, near the Brazen Head, Madeira.

In the islands of Arran, Sky, and other parts
of Scotland, where sandstone, conglomerate, and
other hard rocks are traversed by dikes of trap,
the converse of the above phenomenon is seen.

* I have been favoured with this drawing by Captain
B. Hall.

I

The dike having decomposed more rapidly than the containing rock, has once more left open the original fissure, often for a distance of many yards

Fig. 90.

inland from the sea-coast, as represented in the annexed view. (Fig. 90.) In these instances the greenstone of the dike is usually more tough and hard than the sandstone; but chemical action, and chiefly the oxidation of the iron, has given rise to the more rapid decay.

Fissures left vacant by decomposed trap.
Strathaird, Sky. (MacCulloch.)

There is yet another case, by no means uncommon in Arran and other parts of Scotland, where the strata in contact with the dike, and for a certain distance from it, have been hardened, so as to resist the action of the weather more than the dike itself, or the surrounding rocks. When this happens, two parallel walls of indurated strata are seen protruding above the general level of the country, and following the course of the dike.

As fissures sometimes send off branches, or divide into two or more fissures of equal size, so also we find trap dikes bifurcating and ramifying, and sometimes they are so tortuous as to be called veins, though this is more common in granite than in trap. The accompanying sketch (Fig. 91.) by Dr.

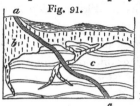

Fig. 91.

MacCulloch represents part of a sea-cliff in Argyleshire, where an overlying mass of trap, *b*, sends out some veins which terminate downwards. Another trap vein, *a a*, cuts

Trap veins in Airdnamurchan.

through both the limestone, *c*, and the trap, *b*.

In Fig. 92. a ground plan is given of a ramifying dike of greenstone, which I observed cutting through sandstone on the beach near Kildonan Castle, in Arran.   The larger branch varies from five to seven feet in width, which will afford a scale of measurement for the whole.

Fig. 92.

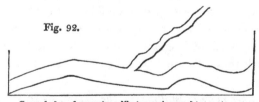

Ground plan of greenstone dike traversing sandstone. Arran.

In the Hebrides and other countries the same masses of trap which occupy the surface of the country far and wide, concealing the subjacent

Fig. 93.

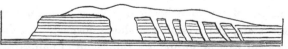

Trap dividing and covering sandstone near Suishnish in Sky. (MacCulloch.)

stratified rocks, are seen also in the sea-cliffs, prolonged downwards in veins or dikes, which probably unite with other masses of igneous rock at
a greater depth. The largest of the dikes represented in the annexed diagram, and which are
seen in part of the coast of Sky, is no less than
100 feet in width.

Every variety of trap rock is sometimes found
in these dikes, as basalt, greenstone, felspar-
porphyry, and more rarely trachyte. The amygdaloidal traps also occur, and even tuff and
breccia, for the materials of these last may be
washed down into open fissures at the bottom of
the sea, or during eruptions on the land may be
showered into them from the air.

Some dikes of trap may be followed for leagues
uninterruptedly in nearly a straight direction, as
in the north of England, showing that the fissures
which they fill must have been of extraordinary
length.

Dikes more crystalline in the centre. — In many
cases trap at the edges or sides of a dike is less
crystalline or more earthy than in the centre, in
consequence of the melted matter having cooled
more rapidly by coming in contact with the cold
sides of the fissure ; whereas, in the centre, the
matter of the dike being kept long in a fluid or
soft state, the crystals are slowly formed. In the
ancient part of Vesuvius a thin band of half-vitre-
ous lava is found at the edge of some dikes. At

the junction of greenstone dikes with limestone, a *sahlband*, or selvage, of serpentine is occasionally observed.

On the left shore of the fiord of Christiania, in Norway, a remarkable dike of syenitic greenstone is traced through transition strata, until at length, in the promontory of Næsodden, it enters mica-schist. Fig. 94. represents a ground plan, where

Fig. 94.

Syenitic greenstone dike of Næsodden, Christiania.

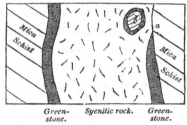

Green-   Syenitic rock.   Greenstone.   stone.

the dike appears eight paces in width. In the middle it is highly crystalline and granitiform, of a purplish colour, and containing a few crystals of mica, and strongly contrasted with the whitish mica-schist, between which and the syenitic rock there is usually on each side a distinct black band, 18 inches wide, of dark greenstone. When first seen, these bands have the appearance of two accompany-ing dikes; yet they are, in fact, only the different form which the syenitic materials have assumed where near to or in contact with the mica-schist. At one point, *a*, one of the sahlbands terminates

for a space; but near this there is a large detached
block *b*, having a gneiss-like structure, consisting
of hornblende and felspar, which is included in the
midst of the dike. Round this a smaller encircling
zone is seen, of dark basalt, or fine-grained greenstone, nearly corresponding to the larger ones
which border the dike, but only one inch wide.*

The fact above alluded to, of a foreign fragment, such as *b* (Fig. 94.), included in the midst
of the trap, as if torn off from some subjacent
rock or the walls of a fissure, is by no means
uncommon. A fine illustration is seen in a dike
of greenstone, ten feet wide, in the northern
suburbs of Christiania, in Norway, of which the
annexed figure is a ground plan. The dike passes

Fig. 95.

Greenstone dike, with fragments of gneiss; Sorgenfri, Christiania.

through shale, known by its fossils to belong to
the transition, or Silurian series. In the black
base of greenstone are angular and roundish
pieces of gneiss, some white, others of a light

* This dike has been described by Professor Keilhau, of
Christiania, in whose company I examined it.

flesh-colour, some without lamination, like granite, others with laminæ, which, by their various and often opposite directions, show that they have been scattered at random through the matrix. These imbedded pieces of gneiss measure from one to about eight inches in diameter.

Rocks altered by volcanic dikes. — After these remarks on the form and composition of dikes themselves, I shall describe the alterations which they sometimes produce in the rocks in contact with them. The changes are usually such as the intense heat of melted matter and the entangled gases might be expected to cause.

Plas-Newydd. — A striking example, near Plas-Newydd, in Anglesea, has been described by Professor Henslow.* The dike is 134 feet wide, and consists of a rock which is a compound of felspar and augite (dolerite of some authors). Strata of shale and argillaceous limestone, through which it cuts perpendicularly, are altered to a distance of thirty, or even, in some places, to thirty-five feet from the edge of the dike. The shale, as it approaches the trap, becomes gradually more compact, and is most indurated where nearest the junction. Here it loses part of its schistose structure, but the separation into parallel layers is still discernible. In several places

* Cambridge Transactions, vol. i. p. 402.

the shale is converted into hard porcellanous jasper. In the most hardened part of the mass the fossil shells, principally *Productæ*, are nearly obliterated; yet even here their impressions may frequently be traced. The argillaceous limestone undergoes analogous mutations, losing its earthy texture as it approaches the dike, and becoming granular and crystalline. But the most extraordinary phenomenon is the appearance in the shale of numerous crystals of analcime and garnet, which are distinctly confined to those portions of the rock affected by the dike.* Garnets have been observed, under very analogous circumstances, in High Teesdale, by Professor Sedgwick, where they occur in shale and limestone, altered by basalt. †

Antrim. — In several parts of the county of Antrim, in the north of Ireland, chalk with flints is traversed by basaltic dikes. The chalk is there converted into granular marble near the basalt, the change sometimes extending eight or ten feet from the wall of the dike, being greatest near the point of contact, and thence gradually decreasing till it becomes evanescent. " The extreme effect," says Dr. Berger, " presents a dark brown crystalline limestone, the crystals running in flakes as large as those of coarse primitive (*metamorphic*)

* Cambridge Transactions, vol. i. p. 410.
† Ibid. vol. ii. p. 175.

limestone; the next state is saccharine, then fine-grained and arenaceous; a compact variety, having a porcellanous aspect and a bluish-grey colour, succeeds: this, towards the outer edge, becomes yellowish-white, and insensibly graduates into the unaltered chalk. The flints in the altered chalk usually assume a grey yellowish colour."* All traces of organic remains are effaced in that part of the limestone which is most crystalline.

The annexed drawing (Fig. 96.) represents three basaltic dikes traversing the chalk, all within

Fig. 96.

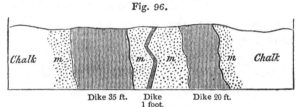

Dike 35 ft.   Dike   Dike 20 ft.
                 1 foot.

Basaltic dikes in chalk in island of Rathlin, Antrim. — Ground plan as seen on the beach. (Conybeare and Buckland.) †

the distance of ninety feet. The chalk contiguous to the two outer dikes is converted into a finely gra-nular marble, *m m*, as are the whole of the masses between the outer dikes and the central one. The entire contrast in the composition and colour of the intrusive and invaded rocks, in these cases, renders the phenomena peculiarly clear and interesting.

Another of the dikes of the north-east of Ire-

* Dr. Berger, Geol. Trans., First Series, vol. iii. p. 172.
† Geol. Trans., First Series, vol. iii. p. 210. and plate 10.

land has converted a mass of red sandstone into hornstone.* By another, the slate clay of the coal measures has been indurated, and has assumed the character of flinty slate †; and in another place the slate clay of the lias has been changed into flinty slate, which still retains numerous impressions of ammonites. ‡

It might have been anticipated that beds of coal would, from their combustible nature, be affected in an extraordinary degree by the contact of melted rock. Accordingly, one of the greenstone dikes of Antrim, on passing through a bed of coal, reduces it to a cinder for the space of nine feet on each side. §

At Cockfield Fell, in the north of England, a similar change is observed. Specimens taken at the distance of about thirty yards from the trap are not distinguishable from ordinary pit coal; those nearer the dike are like cinders, and have all the character of coke; while those close to it are converted into a substance resembling soot. ‖

As examples might be multiplied without end, I shall merely select one or two others, and then conclude. The rock of Stirling Castle is a cal-

* Geol. Trans., First Series, vol. iii. p. 201.

† Ibid. p. 205.

‡ Ibid. p. 213.; and Playfair, Illust. of Hutt. Theory, s. 253.

§ Ibid. p. 206.

‖ Sedgwick, Camb. Trans. vol. ii. p. 37.

careous sandstone, fractured, and forcibly displaced by a mass of greenstone, which has evidently invaded the strata in a melted state. The sandstone has been indurated, and has assumed a texture approaching to hornstone near the junction. In Arthur's Seat and Salisbury Craig, near Edinburgh, a sandstone which comes in contact with greenstone, is converted into a jaspideous rock. *

The secondary sandstones in Sky are converted into solid quartz in several places, where they come in contact with veins or masses of trap; and a bed of quartz, says Dr. MacCulloch, found near a mass of trap, among the coal strata of Fife, was in all probability a stratum of ordinary sandstone, having been subsequently indurated and turned into quartzite by the action of heat.†

But although strata in the neighbourhood of dikes are thus altered in a variety of cases, shale being turned into flinty slate or jasper, limestone into crystalline marble, sandstone into quartz, coal into coke, and the fossil remains of all such strata wholly or in part obliterated, it is by no means uncommon to meet with the same rocks, even in the same districts, absolutely unchanged in the proximity of volcanic dikes.

This great inequality in the effects of the ig-

* Illust. of Hutt. Theory, § 253. and 261. Dr. MacCulloch, Geol. Trans., First Series, vol. ii. p. 305.

† Syst. of Geol. vol. i. p. 206.

neous rocks may often arise from an original difference in their temperature, and in that of the
entangled gases, such as is ascertained to prevail
in different lavas, or in the same lava near its
source and at a distance from it.   The power
also of the invaded rocks to conduct heat may
vary, according to their composition, structure, and
the fractures which they may have experienced, and
perhaps, also, according to the quantity of water
(so capable of being heated) which they contain.
It must happen in some cases that the component
materials are mixed in such proportions as prepare
them readily to enter into chemical union, and
form new minerals; while in other cases the mass
may be more homogeneous, or the proportions less
adapted for such union.

We must also take into consideration, that one
fissure may be simply filled with lava, which may
begin to cool from the first; whereas in other cases
the fissure may give passage to a current of melted
matter, which may ascend for days or months,
feeding streams which are overflowing the country
above, or are ejected in the shape of scoriæ from
some crater. If the walls of a rent, moreover, are
heated by hot vapour before the lava rises, as we
know may happen on the flanks of a volcano, the
additional caloric supplied by the dike and its
gases will act more powerfully.

   Intrusion of trap between strata.—In proof of the

mechanical force which the fluid trap has sometimes exerted on the rocks into which it has intruded itself, I may refer to the Whin-Sill, where
a mass of basalt, from sixty to eighty feet in
height, represented by *a*, Fig. 97., is in part
wedged in between the rocks of limestone, *b*, and
shale, *c*, which have been separated from the great
mass of limestone and shale, *d*, with which they
were united.

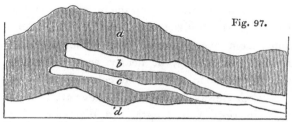

Fig. 97.

*Trap interposed between displaced beds of limestone and shale, at White Force,
High Teesdale, Durham.* (Sedgwick.) *

The shale in this place is indurated; and the
limestone, which at a distance from the trap is
blue, and contains fossil corals, is here converted
into granular marble without fossils.

Masses of trap are not unfrequently met with
intercalated between strata, and maintaining their
parallelism to the planes of stratification throughout large areas. They must in some places have
forced their way laterally between the divisions of
the strata, a direction in which there would be

* Camb. Trans. vol. ii. p. 180.

the least resistance to an advancing fluid, if no
vertical rents communicated with the surface, and
a powerful hydrostatic pressure was caused by
gases propelling the lava upwards.

Columnar and globular structure. — One of the
characteristic forms of volcanic rocks, especially
of basalt, is the columnar, where large masses
are divided into regular prisms, sometimes easily
separable, but in other cases adhering firmly together. The columns vary in the number of angles,
from three to twelve; but they have most commonly from five to seven sides. They are often
divided transversely, at nearly equal distances,
like the joints in a vertebral column, as in the
Giants' Causeway, in Ireland. They vary exceedingly in respect to length and diameter. Dr.
MacCulloch mentions some in Sky which are about
400 feet long; others, in Morven, not exceeding
an inch. In regard to diameter, those of Ailsa
measure nine feet, and those of Morven an inch or
less.* They are usually straight, but sometimes
curved; and examples of both these occur in the
island of Staffa. In a horizontal bed or sheet of
trap the columns are vertical; in a vertical dike
they are horizontal. Among other examples of the
last-mentioned phenomenon is the mass of basalt,
called the Chimney, in St. Helena (see Fig. 98.),
a pile of hexagonal prisms, 64 feet high, evidently

* MacCulloch, Syst. of Geol. vol. ii. p. 137.

Fig. 98.

Volcanic dike composed of horizontal prisms. St. Helena.

Fig. 99.

Small portion of the dike in Fig. 98.

the remainder of a narrow dike, the walls of rock which the dike originally traversed having been removed down to the level of the sea. In Fig. 99. a small portion of this dike is represented on a less reduced scale.*

It being assumed that columnar trap has consolidated from a fluid state, the prisms are said to be always at right angles to the *cooling surfaces*. If these surfaces, therefore, instead of being either perpendicular or horizontal, are curved, the columns ought to be inclined at every angle to the horizon; and there is a beautiful exemplification of this phenomenon in one of the valleys of the Vivarais, a mountainous district in the South of France, where, in the midst of a region of gneiss, a geologist encounters unexpectedly several volcanic cones of loose sand and scoriæ. From the crater of one of these cones, called La Coupe d'Ayzac, a stream of lava descends and occupies the bottom of a narrow valley, except at those points where the river Volant,

* Seale's Geognosy of St. Helena, plate 9.

or the torrents which join it, have cut away portions of the solid lava. The accompanying sketch
(Fig. 100.) represents the remnant of the lava at

Fig. 100.

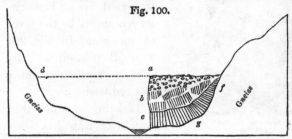

Lava of La Coupe d' Ayzac, near Antraigue, in the Province of Ardêche.

one of the points where a lateral torrent joins the
main valley of the Volant. It is clear that the
lava once filled the whole valley up to the dotted
line *d a*; but the river has gradually swept away
all below that line, while the tributary torrent
has laid open a transverse section; by which we
perceive, in the first place, that the lava is composed, as usual in this country, of three parts; the
uppermost, at *a*, being scoriaceous; the second, *b*,
presenting irregular prisms; and the third, *c*, with
regular columns, which are vertical on the banks
of the Volant, where they rest on a horizontal
base of gneiss, but which are inclined at an angle
of 45° at *g*, and then horizontal at *f*, their position
having been every where determined, according to
the law before mentioned, by the concave form of
the original valley.

In the annexed figure (101.) a view is given

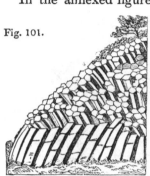

Fig. 101.

of some of the inclined and curved columns which present themselves on the sides of the valleys in the hilly region north of Vicenza, in Italy, and at the foot of the higher Alps.* Unlike those of the Vivarais, last mentioned, the basalt of this country was

Columnar basalt in the Vicentin.
(Fortis.)

evidently submarine, and the present valleys have since been hollowed out by denudation.

The columnar structure is by no means peculiar to the trap rocks in which hornblende or augite predominate; it is also observed in clinkstone, trachyte, and other felspathic rocks of the igneous class, although in these it is rarely exhibited in such regular polygonal forms.

It has been already stated that basaltic columns are often divided by cross joints. Sometimes each segment, instead of an angular, assumes a spheroidal form, so that a pillar is made up of a pile of balls, usually flattened, as in the Cheese-grotto at Bertrich-Baden, in the Eifel, near the

* Fortis, Mém. sur l'Hist. Nat. de l'Italie, tom. i. p. 233. plate 7.

Moselle. (Fig. 102.)　The basalt, there, is part of

Fig. 102.

Basaltic pillars of the Käsegrotte, Bertrich-Baden, half way between Treves and Coblentz.　Height of grotto, from 7 to 8 feet.

a small stream of lava, from 30 to 40 feet thick, which has proceeded from one of several volcanic craters, still extant, on the neighbouring heights. The position of the lava bordering the river in this valley, might be represented by a section like that already given at (Fig. 100. p. 184.), if we merely suppose inclined strata of slate and the argillaceous sandstone called greywacké to be substituted for gneiss.

In some masses of decomposing greenstone, basalt, and other trap rocks, the globular struc-ture is so conspicuous that the rock has the ap-pearance of a heap of large cannon balls.

A striking example of this structure occurs in a resinous trachyte or pitchstone-porphyry in one

of the Ponza islands, which rise from the Mediterranean, off the coast of Terracina and Gaieta.

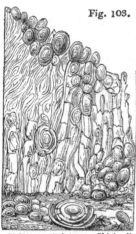

Fig. 103.

The globes vary from a
few inches to three feet
in diameter, and are of
an ellipsoidal form. (See
Fig. 103.) The whole rock
is in a state of decomposition, "and when the balls,"
says Mr. Scrope, " have
been exposed a short time
to the weather, they scale
off at a touch into nu-
merous concentric coats,
like those of a bulbous
root, inclosing a compact
nucleus. The laminæ of

Globiform pitchstone.   Chiaja di
Luna, Isle of Ponza. (Scrope.)

this nucleus have not been so much loosened by
decomposition; but the application of a ruder
blow will produce a still further exfoliation.*

A fissile texture is occasionally assumed by
clinkstone and other trap rocks, so that they have
been used for roofing houses. Sometimes the prismatic and slaty structure is found in the same
mass. The causes which give rise to such arrange-
ments are very obscure, but are supposed to be
connected with changes of temperature during the

* Scrope, Geol. Trans. vol. ii. p. 205. Second Series.

cooling of the mass, as will be pointed out in the
sequel. (See Chap. X.)

Relation of trappean rocks to the products of active volcanos.

When we reflect on the changes above described
in the strata near their contact with trap dikes,
and consider how great is the analogy in composition and structure of the rocks called trappean
and the lavas of active volcanos, it seems difficult
at first to understand how so much doubt could
have prevailed for half a century as to whether
trap was of igneous or aqueous origin. To a certain
extent, however, there was a real distinction between the trappean formations and those to which
the term volcanic was almost exclusively confined.
The trappean rocks first studied in the north of
Germany, and in Norway, France, Scotland, and
other countries, were either such as had been
formed entirely under deep water, or had been injected into fissures and intruded between strata,
and which had never flowed out in the air, or over
the bottom of a shallow sea. When these products,
therefore, of submarine or subterranean igneous
action were contrasted with loose cones of scoriæ,
tuff, and lava, or with narrow streams of lava in
great part scoriaceous and porous, such as were
observed to have proceeded from Vesuvius and
Etna, the resemblance seemed remote and equi-

vocal. It was, in truth, like comparing the roots of
a tree with its leaves and branches, which, although
they belong to the same plant, differ in form, texture, colour, mode of growth, and position. The
external cone, with its loose ashes and porous lava,
may be likened to the light foliage and branches,
and the rocks concealed far below, to the roots.
But it is not enough to say of the volcano,

> " quantum vertice in auras
> " Ætherias, tantum radice in Tartara tendit,"

for its roots do literally reach downwards to Tar-
tarus, or to the regions of subterranean fire; and
what is concealed far below, is probably always
more important in volume and extent than what
is visible above ground.

We have already stated how frequently dense
masses of strata have been removed by denudation
from wide areas (see Chap. VI.) ; and this fact
prepares us to except a similar destruction of whatever may once have formed the uppermost part
of ancient submarine or subaerial volcanos, more
especially as those superficial parts are always of
the lightest and most perishable materials. The
abrupt manner in which dikes of trap usually terminate at the surface (see Fig. 104.), and the water-
worn pebbles of trap in the alluvium which covers
the dike, prove incontestably that whatever was
uppermost in these formations has been swept

Fig. 104.

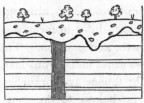

Strata intersected by a trap dike, and covered with alluvium.

away. It is easy, therefore, to conceive that what is gone in regions of trap may have corresponded to what is now visible in active volcanos.

It will be shown in the second part of this volume, that in the earth's crust there are volcanic tuffs of all ages, containing marine shells, which bear witness to eruptions at many successive geological periods. These tuffs, and the associated trappean rocks, must not be compared to lava and scoriæ which had cooled in the open air. Their counterparts must be sought in the products of modern submarine volcanic eruptions. If it be objected that we have no opportunity of studying these last, it may be answered, that subterranean movements have caused, almost everywhere in regions of active volcanos, great changes in the relative level of land and sea, in times comparatively modern, so as to expose to view the effects of volcanic operations at the bottom of the sea.

Thus, for example, the recent examination of the igneous rocks of Sicily, especially those of the Val di Noto, has proved that all the more ordinary varieties of European trap have been there produced under the waters of the sea, at a modern

period; that is to say, since the Mediterranean
has been inhabited by a great proportion of the
existing species of testacea.

These igneous rocks of the Val di Noto, and
the more ancient trappean rocks of Scotland and
other countries, differ from subaerial volcanic
formations in being more compact and heavy, and
in forming sometimes extensive sheets of matter
intercalated between marine strata, and sometimes
stratified conglomerates, of which the rounded pebbles are all trap.   They differ also in the absence
of regular cones and craters, and in the want of
conformity of the lava to the lowest levels of existing valleys.

It is highly probable, however, that insular
cones did exist in some parts of the Val di Noto;
and that they were removed by the waves, in the
same manner as the cone of Graham Island, in
the Mediterranean, was swept away in 1831, and
that of Nyöe, off Iceland, in 1783. All that would
remain in such cases, after the bed of the sea has
been upheaved and laid dry, would be dikes and
shapeless masses of igneous rock, cutting through
sheets of lava, which may have spread over the level
bottom of the sea, and strata of tuff, formed of
materials first scattered far and wide by the winds
and waves, and then deposited.   Trap conglomerates also, to which the action of the waves must
give rise during the denudation of such volcanic

islands, will emerge from the deep whenever the bottom of the sea becomes land. *

The proportion of volcanic matter which is originally submarine must always be very great, as those volcanic vents which are not entirely beneath the sea, are almost all of them in islands, or, if on continents, near the shore. This may explain why extended sheets of trap so often occur, instead of narrow threads, like lava streams. For, a multitude of causes tend, near the land, to reduce the bottom of the sea to a nearly uniform level, — the sediment of rivers, — materials transported by the waves and currents of the sea from wasting cliffs, — showers of sand and scoriæ ejected by volcanos, and scattered by the wind and waves. When, therefore, lava is poured out on such a surface, it will spread far and wide in every direction in a liquid sheet, which may afterwards, when raised up, form the tabular capping of the land.

As to the absence of porosity in the trappean formations, the appearances are in a great degree deceptive, for all amygdaloids are, as already explained, porous rocks, into the cells of which mineral matter, such as silex, carbonate of lime, and other ingredients, have been subsequently introduced. (See p. 156.)

In the little Cumbray, one of the Western

* See Princ. of Geol., *Index*, " Graham Island," " Nyöe," " Conglomerates, volcanic," &c.

Islands, near Arran, the amygdaloid sometimes contains elongated cavities filled with brown spar; and when the nodules have been washed out, the interior of the cavities is glazed with the vitreous varnish so characteristic of the pores of slaggy lavas. Even in some parts of this rock which are excluded from air and water, the cells are empty, and seem to have always remained in this state, and are therefore undistinguishable from some modern lavas.[*]

Dr. MacCulloch, after examining with great attention these and the other igneous rocks of Scotland, observes, " that it is a mere dispute about terms, to refuse to the ancient eruptions of trap the name of submarine volcanos; for they are such in every essential point, although they no longer eject fire and smoke." [†] The same author also considers it not improbable that some of the volcanic rocks of the same country may have been poured out in the open air. [‡]

Although the principal component minerals of subaerial lavas are the same as those of intrusive trap, and both the columnar and globular structure are common to both, there are, nevertheless, some volcanic rocks which never occur as lava, such as greenstone, clinkstone, the more crystalline porphyries, and all those traps in which

[*] MacCulloch, West. Isl., vol. ii. p. 487.
[†] Syst. of Geol., vol. ii. p. 114.   [‡] Ibid.

quartz and mica frequently appear as constituent parts. In short the intrusive trap rocks, forming the intermediate step between lava and the plutonic rocks, depart in their characters from lava in proportion as they approximate to granite.

These views respecting the relations of the volcanic and trap rocks will be better understood, when the reader has studied, in the next chapter, what is said of the plutonic formations.

CHAPTER IX.

PLUTONIC ROCKS — GRANITE.

THE plutonic rocks may be treated of next in order, as they are most nearly allied to the volcanic class, already considered. I have described, in the first chapter, these plutonic rocks as the unstratified division of the crystalline or hypogene formations, and have endeavoured to point out in the Frontispiece, at D, the position which they occupy, when first formed, relatively to the volcanic formations, B.

By some writers, all the rocks now under consideration have been comprehended under the name of granite, which is, then, understood to embrace a large family of crystalline and compound rocks, usually found underlying all other formations; whereas we have seen that trap very commonly overlies strata of different ages. Granite often preserves a very uniform character throughout a wide range of territory, forming hills of a peculiar rounded form, usually clad with a scanty vegetation. The surface of the rock is for the most part in a crumbling state, and the hills are often surmounted by piles of stones like the remains of a stratified mass, as in the annexed figure, and sometimes like heaps of boulders, for

Fig. 105.

Mass of granite near the Sharp Tor, Cornwall.

which they have been mistaken. The exterior of these stones, originally quadrangular, acquires a rounded form by the action of air and water, for the edges and angles waste away more rapidly than the sides. A similar spherical structure has already been described as characteristic of basalt, and other volcanic formations, and it must be referred to analogous causes, as yet but imperfectly understood.

Although it is the general peculiarity of granite
to assume no definite shapes, it is nevertheless
occasionally subdivided by fissures, so as to assume
a cuboidal, and even a columnar, structure.   Examples of these appearances may be seen near the
Land's End, in Cornwall, (see figure.)

Fig. 106.

*Granite having a cuboidal and rude columnar structure.
Land's End, Cornwall.*

The plutonic formations also agree with the
volcanic, in having veins or ramifications proceeding from central masses into the adjoining rocks,
and causing alterations in these last, which will
be presently described.   They also resemble trap

K 3

in containing no organic remains; but they differ in being more uniform in texture, whole mountain masses of indefinite extent appearing to have originated under conditions precisely similar. But they differ in never being scoriaceous or amygdaloidal, in never forming a porphyry with an uncrystalline base, and never alternating with tuffs. Nor do they form conglomerates, although there is sometimes an insensible passage from a fine to a coarse grained granite, and occasionally patches of a fine texture are imbedded in a coarser variety.

Felspar, quartz, and mica are usually considered as the minerals essential to granite, the felspar being most abundant in quantity, and the proportion of quartz exceeding that of mica. These minerals are united in what is termed a confused crystallization; that is to say, there is no regular arrangement of the crystals in granite, as in gneiss (see Fig. 107.), except in the variety

Fig. 107.

Gneiss. (See description, p. 221.)

termed graphic granite, which occurs mostly in

granitic veins. This variety is a compound of felspar and quartz, so arranged as to produce an
imperfect laminar structure. The crystals of felspar

Fig. 108.   Fig. 109.

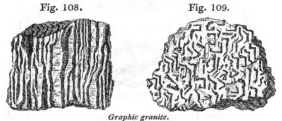

Graphic granite.

Fig. 108.   Section parallel to the laminæ.
Fig. 109.   Section transverse to the laminæ.

appear to have been first formed, leaving between
them the space now occupied by the darker coloured
quartz.   This mineral, when a section is made at
right angles to the alternate plates of felspar and
quartz, presents broken lines, which have been
compared to Hebrew characters.

Porphyritic granite. — This name has been
sometimes given to that variety in which large
crystals of felspar, sometimes more than an inch
in length, are scattered through an ordinary base
of granite.   An example of this texture may be
seen in the granite of the Land's End, in Cornwall. (Fig. 110.) The two larger prismatic crystals
in this drawing represent felspar, smaller crystals
of which are also seen, similar in form, scattered
through the base.   In this base also appear black
specks of mica, the crystals of which have a more

K 4

Fig. 110.

Porphyritic granite, Land's End, Cornwall.

or less perfect hexagonal outline. The remainder
of the mass is quartz, the translucency of which is
strongly contrasted to the opaqueness of the white
felspar and black mica.   But neither this trans-
parency of the quartz, nor the silvery lustre of the
mica, can be expressed in the engraving.

The uniform mineral character of large masses
of granite seems to indicate that large quantities
of the component elements were thoroughly mixed
up together, and then crystallized under precisely
similar conditions.   There are, however, many
accidental, or " occasional, " minerals, as they are
termed, which belong to granite.   Among these
black schorl or tourmaline, actinolite, zircon, garnet, and fluor spar, are not uncommon ; but they
are too sparingly dispersed to modify the general
aspect of the rock.   They show, nevertheless, that
the ingredients were not everywhere exactly the
same ; and a still greater variation may be traced
in the ever-varying proportions of the felspar,
quartz, and mica.

Syenite. — When hornblende is the substitute for mica, which is very commonly the case, the rock becomes Syenite: so called from the celebrated ancient quarries of Syene in Egypt. It has all the appearance of ordinary granite, except when mineralogically examined in hand specimens, and being fully entitled to rank as a geological member of the same plutonic family as granite. Syenite, however, after maintaining the granitic character throughout extensive regions, is not uncommonly found to lose its quartz, and to pass insensibly into syenitic-greenstone, a rock of the trap family.

Syenitic-granite. — The quadruple compound of quartz, felspar, mica, and hornblende, may be so termed. This rock occurs in Scotland and in Guernsey.

Talcose granite, or Protogine of the French, is a mixture of felspar, quartz, and talc. It abounds in the Alps, and in some parts of Cornwall, producing by its decomposition the china clay, more than 12,000 tons of which are annually exported from that county for the potteries.*

Schorl rock, and schorly granite. — The former of these is an aggregate of schorl, or tourmaline, and quartz. When felspar and mica are also present, it may be called schorly granite. This kind of granite is comparatively rare.

* Boase on Primary Geology, p. 16.

K 5

Eurite. — A rock in which all the ingredients
of granite are blended into a finely granular mass.
Crystals of quartz and mica are sometimes scattered through the base of Eurite.

Pegmatite. — A name given by French writers
to a variety of granite; a granular mixture of
quartz and felspar; frequent in granite veins;
passes into graphic granite.

All these granites pass into certain kinds of
trap, a circumstance which affords one of many
arguments in favour of what is now the prevailing
opinion, that the granites are also of igneous
origin.   The contrast of the most crystalline form
of granite, to that of the most common and earthy
trap, is undoubtedly great; but each member of
the volcanic class is capable of becoming porphyritic, and the base of the porphyry may be
more and more crystalline, until the mass passes
to the kind of granite most nearly allied in mineral
composition.

The minerals which constitute alike the granitic
and volcanic rocks, consist, almost exclusively, of
seven elements, namely, silica, alumina, magnesia,
lime, soda, potash, and iron; and these may sometimes exist in about the same proportions in a
porous lava, a compact trap, or a crystalline
granite.   It may perhaps be found, on farther
examination, for on this subject we have yet much
to learn, that the presence of these elements in
certain proportions is more favourable than in

others to their assuming a crystalline or true granitic structure; but it is also ascertained by experiment, that the same materials may, under different circumstances, form very different rocks. The same lava, for example, may be glassy, or scoriaceous, or stony, or porphyritic, according to the more or less rapid rate at which it cools; and some trachytes and syenitic-greenstones may doubtless form granite and syenite, if the crystallization take place slowly.

It would be easy to multiply examples and authorities to prove the gradation of the granitic into the trap rocks. On the western side of the fiord of Christiania, in Norway, there is a large district of trap, chiefly greenstone-porphyry, and syenitic-greenstone, resting on fossiliferous strata. To this, on its southern limit, succeeds a region equally extensive of syenite, the passage from the volcanic to the plutonic rock being so gradual that it is impossible to draw a line of demarcation between them.

" The ordinary granite of Aberdeenshire," says Dr. MacCulloch, " is the usual ternary compound of quartz, felspar, and mica; but sometimes hornblende is substituted for the mica. But in many places a variety occurs which is composed simply of felspar and hornblende; and in examining more minutely this duplicate compound, it is observed in some places to assume a fine grain, and at

length to become undistinguishable from the
greenstones of the trap family. It also passes in
the same uninterrupted manner into a basalt, and
at length into a soft claystone, with a schistose
tendency on exposure, in no respect differing
from those of the trap islands of the western
coast."* The same author mentions, that in
Shetland, a granite composed of hornblende, mica,
felspar, and quartz, graduates in an equally perfect
manner into basalt.†

In Hungary there are varieties of trachyte,
which, geologically speaking, are of modern origin,
in which crystals, not only of mica, but of quartz,
are common, together with felspar and hornblende.
It is easy to conceive how such volcanic masses
may, at a certain depth from the surface, pass
downwards into granite.

I have already hinted at the close analogy in
the forms of certain granitic and trappean veins;
and it will be found that strata penetrated by
plutonic rocks have suffered changes very similar
to those exhibited near the contact of volcanic
dikes. Thus, in Glen Tilt, in Scotland, alternating strata of limestone and argillaceous schist
come in contact with a mass of granite. The
contact does not take place as might have been
looked for, if the granite had been formed there

* Syst. of Geol., vol. i. p. 157.   † Ibid., p. 158.

before the strata were deposited, in which case
the section would have appeared as in Fig. 111.;
but the union is as represented in Fig. 112., the

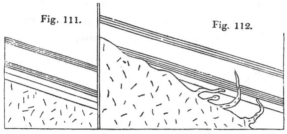

Fig. 111.

Fig. 112.

*Junction of granite and argillaceous schist in Glen
Tilt.* (MacCulloch.) *

undulating outline of the granite intersecting different strata, and occasionally intruding itself in
tortuous veins into the beds of clay-slate and
limestone, from which it differs so remarkably
in composition.   The limestone is sometimes
changed in character by the proximity of the
granitic mass or its veins, and acquires a more
compact texture, like that of hornstone or chert,
with a splintery fracture, effervescing feebly with
acids.

The annexed diagram (Fig. 113.) represents
another junction, in the same district, where the
granite sends forth so many veins as to reticulate
the limestone and schist, the veins diminishing
towards their termination to the thickness of a

* Geol. Trans., First Series, vol. iii. pl. 21.

leaf of paper or a thread. In some places fragments of granite appear entangled, as it were, in
the limestone, and are not visibly connected with
any larger mass; while sometimes, on the other
hand, a lump of the limestone is found in the
midst of the granite. The ordinary colour of

Fig. 113.

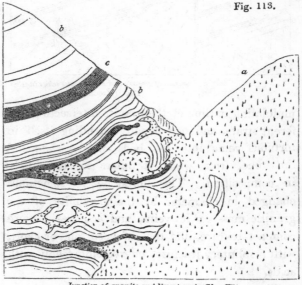

Junction of granite and limestone in Glen Tilt.

a. Granite.   b. Limestone.
c. Blue argillaceous schist.

the limestone of Glen Tilt is lead blue, and its
texture large-grained and highly crystalline; but
where it approximates to the granite, particularly
where it is penetrated by the smaller veins, the

crystalline texture disappears, and it assumes an appearance exactly resembling that of hornstone. The associated argillaceous schist often passes into hornblende slate, where it approaches very near to the granite.*

The conversion of the limestone in these and many other instances into a siliceous rock, effervescing slowly with acids, would be difficult of explanation, were it not ascertained that such limestones are always impure, containing grains of quartz, mica, or felspar disseminated through them. The elements of these minerals, when the rock has been subjected to great heat, may have been fused, and so spread more uniformly through the whole mass.

In the plutonic, as in the volcanic rocks, there is every gradation from a tortuous vein to the most regular form of a dike, such as intersect the tuffs and lavas of Vesuvius and Etna. Dikes of granite may be seen, among other places, on the southern flank of Mount Battock, one of the Grampians, the opposite walls sometimes preserving an exact parallelism for a considerable distance.

As a general rule, however, granite veins in all quarters of the globe are more sinuous in their course than those of trap. They present

* MacCulloch, Geol. Trans., vol. iii. p. 259.

Fig. 114.

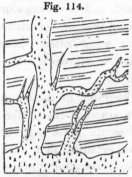

Granite veins traversing clay slate, Table Mountain, Cape of Good Hope.

similar shapes at the most northern point of Scotland, and the southernmost extremity of Africa, as the annexed drawings will show.

It is not uncommon for one set of granite veins to intersect another; and sometimes there are three sets, as in the environs of Heidelberg, where the granite on the banks of the river Necker is seen to consist of three varieties, differing in colour, grain, and various peculiarities of mineral composition. One of these, which is evidently the second in age, is seen to cut through an older granite; and another, still newer, traverses both the second and the first.

In Shetland there are two kinds of granite. One of them, composed of hornblende, mica, felspar, and quartz, is of a dark colour, and is seen underlying gneiss. The other is a red granite, which penetrates the dark variety everywhere in veins.†

The accompanying sketches will explain the

* Capt. B. Hall, Trans. Roy. Soc. Edin., vol. vii.
† MacCulloch, Syst. of Geol., vol. i. p. 58.

manner in which granite veins often ramify and
cut each other. (Figs. 115. and 116.) They re-

Fig. 115.

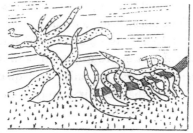

Granite veins traversing gneiss, Cape Wrath. (MacCulloch.) *

present the manner in which the gneiss at Cape
Wrath, in Sutherlandshire, is intersected by veins.
Their light colour, strongly contrasted with that
of the hornblende-schist, here associated with the
gneiss, renders them very conspicuous.

Fig. 116.

Granite veins traversing gneiss at Cape Wrath, in Scotland. (MacCulloch.)

Granite very generally assumes a finer grain,
and undergoes a change in mineral composition,

* Western Islands, pl. 31.

in the veins which it sends into contiguous rocks. Thus, according to Professor Sedgwick, the main body of the Cornish granite is an aggregate of mica, quartz, and felspar; but the veins are sometimes without mica, being a granular aggregate of quartz and felspar. In other varieties quartz prevails to the almost entire exclusion both of felspar and mica; in others, the mica and quartz both disappear, and the vein is simply composed of white granular felspar. *

Fig. 117.

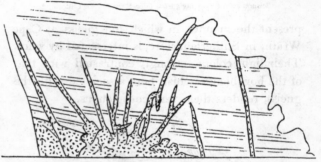

Granite veins passing through hornblende slate, Carnsilver Cove, Cornwall.

Fig. 117. is a sketch of a group of granite veins in Cornwall, given by Messrs. Von Oeynhausen and Von Dechen. † The main body of the granite here is of a porphyritic appearance, with large

* On Geol. of Cornwall, Trans. of Cambridge Soc., vol. i. p. 124.

† Phil. Mag. and Annals, No. 27. New Series, March, 1829.

crystals of felspar; but in the veins it is fine-grained, and without these large crystals. The general height of the veins' is from sixteen to twenty feet, but some are much higher.

In the Valorsine, a valley not far from Mont Blanc, in Switzerland, an ordinary granite, consisting of felspar, quartz, and mica, sends forth veins into a talcose gneiss (or stratified protogine), and in some places lateral ramifications are thrown off from the principal veins at right angles (see Fig. 118.) the veins, especially the minuter ones, being finer grained than the granite in mass.

Fig. 118.

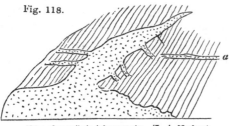

Veins of granite in talcose gneiss. (L. A. Necker.)

It is here remarked, that the schist and granite, as they approach, seem to exercise a reciprocal influence on each other, for both undergo a modification of mineral character. The granite, still remaining unstratified, becomes charged with green particles; and the talcose gneiss assumes a granitiform structure, without losing its stratification. *

* Necker, sur la Val. de Valorsine, Mém. de la Soc. de Phys. de Génève, 1828. — I visited, in 1832, the spot referred to in Fig. 118.

Professor Keilhau drew my attention to several localities in the country near Christiania, where the mineral character of gneiss appears to have been affected by a granite of much newer origin, for some distance from the point of contact. The gneiss, without losing its laminated structure, seems to have become charged with a larger quantity of felspar, and that of a redder colour, than the felspar usually belonging to the gneiss of Norway.

Granite, syenite, and those porphyries which have a granitiform structure, in short all plutonic rocks, are frequently observed to contain metals, at or near their junction with stratified formations. On the other hand, the veins which traverse stratified rocks are, as a general law, more metalliferous near such junctions than in other positions. Hence it has been inferred that these metals may have been spread in a gaseous form through the fused mass, and that the contact of another rock, in a different state of temperature, or sometimes the existence of rents in other rocks in the vicinity, may have caused the sublimation of the metals. *

There are many instances, as at Markerud, near Christiania, in Norway, where the strike of the beds has not been deranged throughout a large area by the intrusion of granite, both in large masses and in veins.   This fact is considered by

* Necker, Proceedings of Geol. Soc., No. 26. p. 392.

some geologists to militate against the theory of
the forcible injection of granite in a fluid state.
But it may be stated in reply, that ramifying
dikes of trap, which almost all now admit to have
been once fluid, pass through the same fossiliferous strata, near Christiania, without deranging
their strike or dip. *

The real or apparent isolation of large or small
masses of granite detached from the main body, as at
a b, Fig.119., and above, Fig.113., and *a*, Fig.118.,

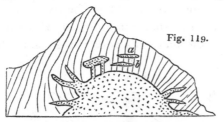

Fig. 119.

General view of junction of granite and schist of the Valorsine.
(L. A. Necker.)

has been thought by some writers to be irrecon-
cilable with the doctrine usually taught respecting veins; but many of them may, in fact, be
sections of root-shaped prolongations of granite;
while, in other cases, they may in reality be detached portions of rock having the plutonic structure. For there may have been spots in the midst
of the invaded strata, in which there was an assemblage of materials more fusible than the rest,

* See Keilhau's Gæa Norvegica; Christiania, 1838.

or more fitted to combine readily into some form
of granite.

Veins of pure quartz are often found in granite,
as in many stratified rocks, but they are not traceable, like veins of granite or trap, to large bodies
of rock of similar composition. They appear to
have been cracks, into which siliceous matter was
infiltered. Such segregation, as it is called, can
sometimes be shown to have clearly taken place
long subsequently to the original consolidation of
the containing rock. Thus, for example, in the
gneiss of Tronstad Strand, near Drammen, in
Norway, the annexed section is seen on the beach.

Fig. 120.

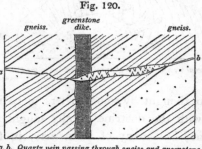

*a. b. Quartz vein passing through gneiss and greenstone,
Tronstad Strand, near Christiania.*

It appears that the alternating strata of whitish
granitiform gneiss, and black hornblende-schist,
were first cut through by a greenstone dike, about
$2\frac{1}{2}$ feet wide; then the crack *a b* passed through
all these rocks, and was filled up with quartz.
The opposite walls of the vein are in some parts

incrusted with transparent crystals of quartz, the middle of the vein being filled up with common opaque white quartz.

We have seen that the volcanic formations have been called overlying, because they not only penetrate others, but spread over them.   Mr. Necker has proposed to call the granites the underlying igneous rocks, and the distinction here indicated is highly characteristic.   It was indeed supposed by Von Buch, at the commencement of his geological career, that the granite of Christiania, in Norway, was sometimes intercalated in mountain masses between the transition strata of that country, overlying fossiliferous shale and limestone. But although the granite sends veins into these fossiliferous rocks, and is decidedly posterior in origin, the opinion expressed of its actual superposition in mass has been disproved by Professor Keilhau, some of whose observations respecting localities referred to by Von Buch, I have lately had opportunities of verifying.   There are, however, on a smaller scale, certain beds of euritic porphyry, some a few feet, others many yards in thickness, which pass into granite, and deserve perhaps to be classed as plutonic rather than trappean rocks, which may truly be described as interposed conformably between fossiliferous strata, as the porphyries ($a\ c$, Fig. 121.), which divide the bituminous shales and argillaceous limestones, $f\ f$.

Fig. 121.

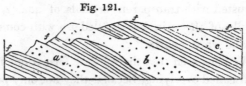

*Euritic porphyry alternating with fossiliferous transition strata,
near Christiania.*

But some of these same porphyries are partially
unconformable, as *b*, and may lead us to suspect
that the others also, notwithstanding their appearance of interstratification, have been forcibly injected.   Some of the porphyritic rocks above
mentioned are highly quartzose, others very felspathic.   In proportion as the masses are more
voluminous, they become more granitic in their
texture, less conformable, and even begin to send
forth veins into contiguous strata.   In a word, we
have here a beautiful illustration of the intermediate gradations between volcanic and plutonic
rocks, not only in their mineralogical composition
and structure, but also in their relations of position to associated formations.   If the term overlying can in this instance be applied to a plutonic
rock, it is only in proportion as that rock begins
to acquire a trappean aspect.

It has been already hinted that the heat, which
in every active volcano extends downwards to indefinite depths, must produce simultaneously very
different effects near the surface, and far below it;
and we cannot suppose that rocks resulting from

the crystallizing of fused matter under a pressure of several miles of the earth's crust can resemble those formed at or near the surface.   Hence the production at great depths of a class of rocks analogous to the volcanic, and yet differing in many particulars, might almost have been predicted, even had we no plutonic formations to account for.   How well these agree, both in their positive and negative characters with the theory of their deep subterranean origin, the student will be able to judge by considering the descriptions already given.

It has, however, been objected, that if the granitic and volcanic rocks were simply different parts of one great series, we ought to find in mountain chains volcanic dikes passing upwards into lava, and downwards into granite.   But we may answer, that our vertical sections are usually of small extent; and if we find in certain places a transition from trap to porous lava, and in others a passage from granite to trap, it is as much as could be expected of this evidence.

The prodigious extent of denudation which has been already demonstrated to have occurred at former periods, will reconcile the student to the belief, that crystalline rocks of high antiquity, although deep in the earth's crust when originally formed, may have become uncovered and exposed at the surface.   Their actual elevation above the

sea may be referred to the same causes to which
we have attributed the upheaval of marine strata,
even to the summits of some mountain chains.
But to these and other topics, I shall revert when
speaking, in the second part, of the relative ages
of different masses of granite.

CHAPTER X.

METAMORPHIC ROCKS.

General character of metamorphic rocks — Gneiss — Horn-blende-schist — Mica-schist — Clay-slate — Quartzite — Chlorite-schist — Metamorphic limestone — Alphabetical list and explanation of other rocks of this family — Origin of the metamorphic strata — Their stratification is real and distinct from cleavage — On joints and slaty cleavage — Supposed causes of these structures — how far connected with crystalline action.

WE have now considered three distinct classes of rocks: first, the aqueous, or fossiliferous; secondly, the volcanic; and, thirdly, the plutonic, or granitic; and we have now lastly to examine those crystalline strata to which the name of *metamorphic* has been assigned. The last-mentioned term expresses, as before explained, a theoretical opinion that such strata, after having been deposited from water, acquired by the influence of heat and other causes a highly crystalline texture.

These rocks, when in their most characteristic or normal state, are wholly devoid of organic remains, and contain no distinct fragments of other rocks whether rounded or angular. They sometimes break out in the central parts of narrow mountain chains, but in other cases extend over areas of vast dimensions, occupying, for example, nearly the whole of Norway and Sweden, where, as in Brazil, they appear alike in the lower and higher grounds. In Great Britain those members of the series which approach most nearly to granite in their composition, as gneiss, mica-schist and hornblende-schist, are confined to the country north of the rivers Forth and Clyde.

Many attempts have been made to trace a general order of succession or superposition in the members of this family; gneiss, for example, having been often supposed to hold invariably a lower geological position than mica-schist. But although such an order may prevail throughout limited districts, it is by no means universal, nor even general throughout the globe. To this subject, however, we shall again revert in the second part of this volume, when the chronological relations of the metamorphic rocks are pointed out.

The following may be enumerated as the principal members of the metamorphic class, gneiss, mica-schist, hornblende-schist, clay-slate, chlorite-schist, hypogene or metamorphic limestone, and certain kinds of quartz rock or quartzite.

Gneiss. — The first of these, gneiss, may be called stratified granite, being formed of the same materials as granite, namely felspar, quartz, and mica. In the specimen here figured, the

Fig. 122.

Fragment of gneiss, natural size, section at right angles to planes of stratification.

white layers consist almost exclusively of granular felspar, with here and there a speck of mica and grain of quartz. The dark layers are composed of grey quartz and black mica, with occasionally a grain of felspar intermixed. The rock splits most easily in the plane of these darker layers, and the surface thus exposed is almost entirely covered with shining spangles of mica. The accompanying quartz however greatly predominates in quantity, but the most ready cleavage is determined by the abundance of mica in certain parts of the dark layer.

Instead of these thin laminæ gneiss is sometimes simply divided into thick beds, in which the mica has only a slight degree of parallelism to the planes of stratification.

The term " gneiss," however, in geology is commonly used in a wider sense to designate a formation in which the above mentioned rock prevails, but with which any one of the other

metamorphic rocks, and more especially hornblende-schist, may alternate.   These other members of the metamorphic series are, in this case,
considered as subordinate to the true gneiss.   In
some rare instances fragments of pre-existing
rocks may be detected in gneiss.

The different varieties of rock allied to gneiss,
into which felspar enters as an essential ingredient, will be understood by referring to what was
said of granite.   Thus, for example, hornblende
may be superadded to mica, quartz, and felspar,
forming a syenitic gneiss; or talc may be substituted for mica, constituting talcose gneiss, a
rock composed of felspar, quartz, and talc, in
distinct crystals or grains (stratified protogine of
the French).

Hornblende-schist is usually black, and composed
principally of hornblende, with a variable quantity of felspar, and sometimes grains of quartz.
When the hornblende and felspar are nearly in
equal quantities, and the rock is not slaty, it cor-
responds in character with the greenstones of the
trap family, and has been called " primitive greenstone."   Some of these hornblendic masses may
really have been volcanic rocks, which have since
assumed a more crystalline or metamorphic texture.

Mica-schist, or *micaceous schist,* is, next to gneiss,
one of the most abundant rocks of the metamor-

phic series.   It is slaty, essentially composed of
mica and quartz, the mica sometimes appearing
to constitute the whole mass.   Beds of pure
quartz also occur in this formation.   In some districts garnets in regular twelve-sided crystals form
an integrant part of mica-schist.   This rock passes
by insensible gradations into clay-slate.

Clay-slate, or Argillaceous schist. — This rock resembles an indurated clay or shale, is for the most
part extremely fissile, often affording good roofing
slate.   It may consist of the ingredients of gneiss,
or of an extremely fine mixture of mica and
quartz, or talc and quartz.   Occasionally it derives a shining and silky lustre from the minute
particles of mica or talc which it contains.   It
varies from greenish or bluish-grey to a lead
colour.   It may be said of this, more than of any
other schist, that it is common to the metamorphic
and fossiliferous series, for some clay-slates taken
from each division would not be distinguishable
by mineralogical characters.

Quartzite, or Quartz rock, is an aggregate of
grains of quartz, which are either in minute crystals, or in many cases slightly rounded, occurring
in regular strata associated with gneiss or other
metamorphic rocks.   Compact quartz, like that
so frequently found in veins is also found together
with granular quartzite.   Both of these alternate

with gneiss or mica-schist, or pass into those rocks by the addition of mica, or of felspar and mica.

Chlorite-schist is a green slaty rock in which chlorite is abundant in foliated plates, usually blended with minute grains of quartz, or sometimes with felspar or mica.   Often associated with, and graduating into, gneiss and clay-slate.

Hypogene or metamorphic limestone. — This rock, commonly called *primary limestone*, is sometimes a thick bedded white crystalline granular marble used in sculpture ; but more frequently it occurs in thin beds, forming a foliated schist much resembling in colour and appearance certain varieties of gneiss and mica-schist.   It alternates with both these rocks, and in like manner with argillaceous schist.   It then usually contains some crystals of mica, and occasionally quartz, felspar, hornblende, and talc.   This member of the metamorphic series enters sparingly into the structure of the hypogene districts of Norway, Sweden, and Scotland, but is largely developed in the Alps.

Before offering any further observations on the probable origin of the metamorphic rocks, I subjoin in the form of a glossary, a brief explanation of some of the principal varieties and their synonymes.

ACTINOLITE-SCHIST.   A slaty foliated rock, composed chiefly of actinolite, (an emerald-green mineral, allied to hornblende,) with some admixture of felspar, or quartz, or miea.

AMPELITE.   Aluminous slate (Brongniart) ; occurs both in the metamorphic and fossiliferous series.

AMPHIBOLITE.   Hornblende rock, which see.

ARGILLACEOUS-SCHIST, or CLAY-SLATE.   *See* p. 223.

CHIASTOLITE-SLATE scarcely differs from clay-slate, but includes numerous crystals of Chiastolite; in considerable thickness in Cumberland.   Chiastolite occurs in long slender rhomboidal crystals.   For composition, see Table, p. 166.

CHLORITE-SCHIST.   A green slaty rock, in which chlorite, a green scaly mineral, is abundant.   *See* p. 224.

CLAY-SLATE, or ARGILLACEOUS-SCHIST.   *See* p. 223.

EURITE and EURITIC PORPHYRY.   A base of compact felspar, with grains of laminar felspar, and often mica and other minerals disseminated (Brongniart).   M. D'Aubuisson regards eurite as an extremely fine grained granite, in which felspar predominates, the whole forming an apparently homogeneous rock.   Eurite has been already mentioned as a plutonic rock, but occurs also in beds subordinate to gneiss or mica-slate.

GNEISS.   A stratified or laminated rock, same composition as granite.   *See* p. 220.

HORNBLENDE ROCK, or AMPHIBOLITE.   The same composition as hornblende schist, stratified, but not fissile.   *See* p. 163.

HORNBLENDE-SCHIST, or SLATE.   Composed chiefly of hornblende, with occasionally some felspar.   *See* p. 222.

HORNBLENDIC or SYENITIC GNEISS.   Composed of felspar, quartz, and hornblende.

HYPOGENE LIMESTONE.   *See* p. 224.

MARBLE.   *See* p. 224.

MICA-SCHIST, or MICACEOUS-SCHIST.   A slaty rock, composed of mica and quartz in variable proportions.   *See* p. 222.

MICA-SLATE.   *See* MICA-SCHIST, p. 222.

PHYLLADE.   D'Aubuisson's term for clay-slate, from φυλλας, a heap of leaves.

PRIMARY LIMESTONE.   *See* HYPOGENE LIMESTONE, p. 224.

PROTOGINE. *See* TALCOSE-GNEISS, p. 222.: when unstratified it is
    Talcose-granite.

QUARTZ ROCK, or QUARTZITE. A stratified rock; an aggregate
    of grains of quartz. *See* p. 223.

SERPENTINE occurs in both divisions of the hypogene series, as
    a stratified or unstratified rock; contains much magnesia;
    is chiefly composed of the mineral called serpentine, mixed
    with diallage, talc, and steatite. The pure varieties of this
    rock, called noble serpentine, consist of a hydrated silicate
    of magnesia, generally of a greenish colour; this base is
    commonly mixed with oxide of iron.

TALCOSE-GNEISS. Same composition as talcose granite or pro-
    togine, but either stratified or laminated.
TALCOSE-SCHIST consists chiefly of talc, or of talc and quartz,
    or of talc and felspar, and has a texture something like that
    of clay-slate.

WHITESTONE. Same as Eurite.

Origin of the Metamorphic Strata.

Having said thus much of the mineral composition of the metamorphic rocks, I may combine
what remains to be said of their structure and
history, with an account of the opinions entertained of their probable origin. At the same
time it may be well to forewarn the reader that
we are here entering upon ground of controversy,
and soon reach the limits where positive induction
ends, and beyond which we can only indulge in
speculations. It was once a favourite doctrine, and
is still maintained by many, that these rocks owe

their crystalline texture, their want of all signs of
a mechanical origin, or of fossil contents, to a
peculiar and nascent condition of the planet at
the period of their formation. The arguments
in refutation of this hypothesis will be more fully
considered when I show, in the second part of
this volume, to how many different ages the
metamorphic formations are referable, and how
gneiss, mica-schist, clay-slate, and hypogene limestone (that of Carrara for example), have been
formed, not only since the first introduction of
organic beings into this planet, but even long
after many distinct races of plants and animals
had passed away in succession.

The doctrine respecting the crystalline strata,
implied in the name metamorphic, may properly
be treated of in this place; and we must first inquire whether these rocks are really entitled to
be called stratified in the strict sense of having
been originally deposited as sediment from water.
The general adoption by geologists of the term
stratified, as applied to these rocks, sufficiently
attests their division into beds very analogous, at
least in form, to ordinary fossiliferous strata. This
resemblance is by no means confined to the existence in both of an occasional slaty structure, but
extends to every kind of arrangement which is
compatible with the absence of fossils, and of sand,
pebbles, ripple mark, and other characters which

the metamorphic theory supposes to have been obliterated by plutonic action. Thus, for example, we behold alike in the crystalline and fossiliferous formations an alternation of beds varying greatly in composition, colour, and thickness. We observe, for instance, gneiss alternating with layers of black hornblende-schist, or with granular quartz, or limestone; and the interchange of these different strata may be repeated for an indefinite number of times. In the like manner, mica-schist alternates with chlorite-schist, and with granular limestone in thin layers.

As in fossiliferous formations strata of pure siliceous sand alternate with micaceous sand and with layers of clay, so in the crystalline or metamorphic rocks we have beds of pure quartzite alternating with mica-schist and clay-slate. As in the secondary and tertiary series we meet with limestone alternating again and again with micaceous or argillaceous sand, so we find in the hypogene, gneiss and mica-schist alternating with pure and impure granular limestones.

It has also been shown that the ripple mark is very commonly repeated throughout a considerable thickness of fossiliferous strata, so in mica-schist and gneiss, there is sometimes an undulation of the laminæ on a minute scale, which may, perhaps, be a modification of similar inequalities in the original deposit.

In the crystalline formations also, as in many
of the sedimentary before described, single strata
are sometimes made up of laminæ placed diago-
nally, such laminæ not being regularly parallel to
the planes of cleavage.

This disposition of the layers is illustrated in
the accompanying diagram, in which I have re-

Fig. 123.

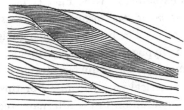

*Lamination of clay-slate, Montagne de Seguinat, near Gavarnie,
in the Pyrenees.*

presented carefully the stratification of a coarse
argillaceous schist, which I examined in the
Pyrenees, part of which approaches in character
to a green and blue roofing slate, while part is
extremely quartzose, the whole mass passing downwards into micaceous schist.   The vertical section
here exhibited is about three feet in height, and
the layers are sometimes so thin that fifty may be
counted in the thickness of an inch.   Some of
them consist of pure quartz.

The inference drawn from the phenomena
above described, in favour of the aqueous origin
of clay-slate and other crystalline strata, is greatly

strengthened by the fact that many of these metamorphic rocks occasionally alternate with, and
sometimes pass, by intermediate gradations, into
rocks of a decidedly mechanical origin, and exhibiting traces of organic remains. The fossiliferous formations, moreover, into which this passage is effected, are by no means invariably of the
same age nor of the highest antiquity, as will be
afterwards explained. (See Part II.)

*Stratification of the metamorphic rocks distinct
from cleavage.* — The beds into which gneiss, mica-
schist, and hypogene limestone divide, exhibit
most commonly, like ordinary strata, a want of
perfect geometrical parallelism. For this reason,
therefore, in addition to the alternate recurrence
of layers of distinct materials, the stratified arrangement of the crystalline rocks cannot be explained away by supposing it to be simply a
divisional structure like that to which we owe
some of the slates used for writing and roofing.
Slaty cleavage as it has been called, has in many
cases been produced by the regular deposition of
thin plates of fine sediment one upon another,
but there are many instances where it is decidedly
unconnected with such a mode of origin, and
where it is not even confined to the aqueous formations. Some kinds of trap, for example, as clinkstone, split into laminæ, and are used for roofing.

There are, says Professor Sedgwick, three distinct forms of structure exhibited in certain rocks throughout large districts: viz. — First, stratification; secondly, joints; and thirdly, slaty cleavage; the two last having no connection with true bedding, and having been superinduced by causes absolutely independent of gravitation. All these different structures must have different names, even though there be some cases where it is impossible, after carefully studying the appearances, to decide upon the class to which they belong. *

Joints. — Now in regard to the second of these forms of structure or joints, they are natural fissures which often traverse rocks in straight and well determined lines. They afford to the quarryman, as Mr. Murchison observes, when speaking of the phenomena, as exhibited in Shropshire and the neighbouring counties, the greatest aid in the extraction of blocks of stone, and, if a sufficient number cross each other, the whole mass of rock is split into symmetrical blocks. † The faces of the joints are for the most part smoother and more regular than the surfaces of true strata. The joints are straight-cut chinks, often slightly open, often passing, not only through layers of successive deposition, but also through balls of limestone or other matter which have been formed

* Geol. Trans., Second Series, vol. iii. p. 480.
† The Silurian System of Rocks, as developed in Salop, Hereford, &c., p. 245.

by concretionary action, since the original accumulation of the strata.   Such joints, therefore,
must often have resulted from one of the last
changes superinduced upon sedimentary deposits.*

In the annexed diagram the flat surfaces of
rock A, B, C, represent exposed faces of joints, to
which the walls of other joints, J J, are parallel.
S S are the lines of stratification ; C C are lines
of slaty cleavage, which intersect the rock at a considerable angle to the planes of stratification.

Fig. 124.

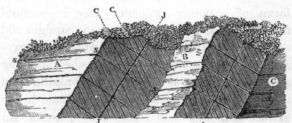

Stratification, joints, and cleavage.

Joints according to Professor Sedgwick are distinguishable from lines of slaty cleavage in this,
that the rock intervening between two joints has
no tendency to cleave in a direction parallel to
the planes of the joints, whereas a rock is capable
of indefinite subdivision in the direction of its
slaty cleavage.   In some cases where the strata
are curved, the planes of cleavage are still per-

* The Silurian System of Rocks, as developed in Salop,
Hereford, &c., p. 246.

fectly parallel. This has been observed in the
slate rocks of part of Wales. (See Fig. 125.) which

Fig. 125.

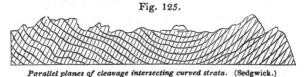

Parallel planes of cleavage intersecting curved strata. (Sedgwick.)

consist of a hard greenish slate. The true bedding is there indicated by a number of parallel
stripes, some of a lighter and some of a darker
colour than the general mass. Such stripes are
found to be parallel to the true planes of stratification, wherever these are manifested by ripplemark,
or by beds containing peculiar organic remains.
Some of the contorted strata are of a coarse mechanical structure, alternating with fine-grained
crystalline chloritic slates, in which case the same
slaty cleavage extends through the coarser and
finer beds, though it is brought out in greater
perfection in proportion as the materials of the
rock are fine and homogeneous. It is only when
these are very coarse that the cleavage planes
entirely vanish. These planes are usually inclined
at a very considerable angle to the planes of the
strata. In the Welsh chains, for example, the average angle is as much as from 30° to 40°. Sometimes the cleavage planes dip towards the same
point of the compass as those of stratification, but
more frequently to opposite points. It may be

stated as a general rule, that when beds of coarser materials alternate with those composed of finer particles, the slaty cleavage is either entirely confined to the fine-grained rock, or is very imperfectly exhibited in that of coarser texture. This rule holds, whether the cleavage is parallel to the planes of stratification or not.

In the Swiss and Savoy Alps, as Mr. Bakewell has remarked, enormous masses of limestone are cut through so regularly by nearly vertical partings, and these are often so much more conspicuous than the seams of stratification, that an unexperienced observer will almost inevitably confound them, and suppose the strata to be perpendicular in places where in fact they are almost horizontal. *

Now these joints are supposed to be analogous to those partings which have been already observed to separate volcanic and plutonic rocks into cuboidal and prismatic masses. On a small scale we see clay and starch when dry split into similar shapes, which is often caused by simple contraction, whether the shrinking be due to the evaporation of water, or to a change of temperature. It is well known that many sandstones and other rocks expand by the application of moderate degrees of heat, and then contract again

* Introduction to Geology, chap. iv.

on cooling; and there can be no doubt that large
portions of the earth's crust have, in the course of
past ages, been subjected again and again to very
different degrees of heat and cold.   These alternations of temperature have probably contributed
largely to the production of joints in rocks.

In some countries, as in Saxony, where masses
of basalt rest on sandstone, the aqueous rock has
for the distance of several feet from the point of
junction assumed a columnar structure similar
to that of the trap.   In like manner some hearth-
stones, after exposure to the heat of a furnace
without being melted, have become prismatic.
Certain crystals also acquire by the application of
heat a new internal arrangement, so as to break
in a new direction, their external form remaining
unaltered.

Scoresby, when speaking of the icebergs of Spitz-
bergen, states that " they are full of rents, extending perpendicularly downwards, and dividing them
into innumerable columns."   Colonel Jackson, who
has lately investigated this subject more attentively, found that the ice on the Neva, at St. Peters-
burg, at the beginning of a thaw, when two feet
in thickness, is traversed by rows of very minute
air-bubbles extending in straight lines, sometimes
a little inflected, from the upper surface of the ice
towards the lower, within from two to five inches
of which they terminate.   " Other blocks pre-

sented these bubbles united, so as to form cylindrical canals, a little thicker than a horsehair.
Observing still further," he says, " I found blocks
in which the process was more advanced, and two,
three, or more clefts, struck off in different directions from the vertical veins, so that a section perpendicular to the vein would represent in miniature
the star-formed cracks of timber.   Finally, in some
pieces, these cracks united from top to bottom of
the veins, separating the whole mass into vertical
prisms, having a greater or less number of sides.
In this state a slight shock was sufficient to detach
them; and the block with its scattered fragments
was in all respects the exact miniature resemblance,
in crystal, of a Giant's Causeway.   The surface
was like a tessellated pavement, and the columns
rose close, adhering and parallel, from the compact mass of a few inches at the under surface.
More or less time is required for the process,
which I have since seen in all its different stages." *

Here again we find the columnar or jointed
structure in a solid mass, which had been subjected to great changes of temperature.

It seems, therefore, that the fissures called joints
may have been the result of different causes, as of
some modification of crystalline action, or simple
contraction during consolidation, or during a

* Journ. of Roy. Geograph. Soc., vol. v. p. 19.

change of temperature. And there are cases where joints may have been due to mechanical violence, and the strain exerted on strata during their upheaval, or when they have sunk down below their former level. Professor Phillips has suggested that the previous existence of divisional planes may often have determined, and must greatly have modified, the lines and points of fracture caused in rocks by those forces to which they owe their elevation or dislocations. These lines and points being those of least resistance, cannot fail to have influenced the direction in which the solid mass would give way on the application of external force.

It has been observed by Mr. Murchison, that in referring both joints and slaty cleavage to crystalline action, we are borne out by a well-known analogy in which crystallization has in like manner given rise to two distinct kinds of structure in the same body. Thus for example, in a six-sided prism of quartz, the planes of cleavage are distinct from those of the prism. It is impossible to cleave the crystals parallel to the plane of the prism, just as slaty rocks cannot be cleaved parallel to the joints, but the quartz crystal, like the older schists, may be cleaved *ad infinitum* in the direction of the cleavage planes. *

* Silurian System of Rocks, &c., p. 246.

I have already stated that extremely fine slates, like those of the Niesen, near the Lake of Thun, in Switzerland, are perfectly parallel to the planes of stratification, and are, therefore, probably due to successive aqueous deposition. Even when the slates are oblique to the general planes of the strata, it by no means follows as a matter of course, that they have been caused by crystalline action, for they may be the result of that diagonal lamination which I have before described (p. 38.). In this case, however, there is usually much irregularity, whereas those cleavage planes oblique to the true stratification, which are referred to a crystalline action, are often perfectly symmetrical, and observe a strict geometrical parallelism, even when the strata are contorted, as already described (p. 233.).

In regard to the origin of slaty cleavage, where it is unconnected with sedimentary deposition, Professor Sedgwick is of opinion that no retreat of parts, no contraction in dimensions, in passing to a solid state, can account for the phenomenon. It must be referred to crystalline or polar forces acting simultaneously and somewhat uniformly, in given directions, on large masses having a homogeneous composition.

A fact recorded by Mr. Darwin affords confirmation to this theory. The ore of the gold mines of Yaquil, in Chili, is ground in a mill into

an impalpable powder. After this powder has
been washed, and nearly all the metal separated,
the mud which passes from the mills is collected
into pools, where it subsides, and is cleared out
and thrown into a common heap. A great deal
of chemical action then commences, salts of various kinds effloresce on the surface, and the mass
becomes hard, and divides into concretionary fragments. These fragments were observed to possess
an even and well defined slaty structure; but the
laminæ were not inclined at any uniform angle.*

Mr. R. W. Fox lately submitted a mass of moist
clay, worked up with acidulated water, to weak
voltaic action for some months, and it was found
when dry to be rudely laminated, the planes of
the slightly undulating laminæ being at right
angles to the direction of the electrical forces.†

Sir John Herschel, in allusion to slaty cleavage,
has suggested, " that if rocks have been so heated
as to allow a commencement of crystallization;
that is to say, if they have been heated to a point
at which the particles can begin to move amongst
themselves, or at least on their own axes, some
general law must then determine the position in
which these particles will rest on cooling. Pro-

* Journal, p. 324. (for title, see note p. 137.).
† Although the lamination in the specimen shown to me
was very imperfect, it was sufficiently evident to encourage
farther experiments.

bably, that position will have some relation to the direction in which the heat escapes. Now, when all, or a majority of particles of the same nature have a general tendency to one position, that must of course determine a cleavage plane. Thus we see the infinitesimal crystals of fresh precipitated sulphate of baryte, and some other such bodies, arrange themselves alike in the fluid in which they float; so as, when stirred, all to glance with one light, and give the appearance of silky filaments. Some sorts of soap, in which insoluble margarates * exist, exhibit the same phenomenon when mixed with water; and what occurs in our experiments on a minute scale may occur in nature on a great one." †

* Margaric acid is an oleaginous acid, formed from different animal and vegetable fatty substances. A margarate is a compound of this acid with soda, potash, or some other base, and is so named from its pearly lustre.

† Letter to the author, dated Cape of Good Hope, Feb. 20. 1836.

CHAPTER XI.

IT has been seen that geologists have been very generally led to infer, from the phenomena of joints and slaty cleavage, that mountain masses, of which the sedimentary origin is unquestionable, have been acted upon simultaneously by vast crystalline forces. That the structure of fossiliferous strata has often been modified by some general cause since their original deposition, and even subsequently to their consolidation and dislocation, is undeniable. These facts prepare us to believe, that still greater changes may have been worked out by a greater intensity, or more prolonged development of the same agency, combined, perhaps, with other causes. Now we have seen that, near the immediate contact of granitic veins and volcanic dikes, very extraordinary alterations in rocks have taken place, more especially

M

in the neighbourhood of granite.   It will be useful here to add other illustrations, showing that
a texture undistinguishable from that which cha-
racterizes the more crystalline metamorphic formations, has actually been superinduced in strata
once fossiliferous.

In the southern extremity of Norway, there is a
large district, on the west side of the fiord of Christiania, in which granite or syenite protrudes in
mountain masses through fossiliferous strata, and
usually sends veins into them at the point of contact.   The stratified rocks, replete with shells and
zoophytes, consist chiefly of shale, limestone, and
some sandstone, and all these are invariably altered
near the granite for a distance of from 50 to 400
yards.   The aluminous shales are hardened and have
become flinty.   Sometimes they resemble jasper.
Ribboned jasper is produced by the hardening of
alternate layers of green and chocolate-coloured
schist, each stripe faithfully representing the original lines of stratification.   Nearer the granite
the schist often contains crystals of hornblende,
which are even met with in some places for a distance of several hundred yards from the junction;
and this black hornblende is so abundant, that
eminent geologists, when passing through the
country, have confounded it with the ancient
hornblende-schist, subordinate to the great gneiss
formation of Norway.   Frequently, between the

granite and the hornblendic slate, above mentioned, grains of mica and crystalline felspar appear in the schist, so that rocks resembling gneiss
and mica-schist are produced.   Fossils can rarely
be detected in these schists, and they are more
completely effaced in proportion to the more crystalline texture of the beds, and their vicinity to
the granite.   In some places the siliceous matter
of the schist becomes a granular quartz, and when
hornblende and mica are added, the altered rock
loses its stratification, and passes into a kind of
granite.   The limestone, which at points remote
from the granite is of an earthy texture, blue
colour, and often abounds in corals, becomes a
white granular marble near the granite, sometimes
siliceous, the granular structure extending occa-

Fig. 126.

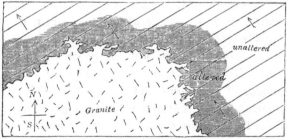

Altered zone of fossiliferous slate and limestone near granite.   Christiania.
The arrows indicate the dip, and the straight lines the strike,
of the beds.

sionally upwards of 400 yards from the junction;
and the corals being for the most part obliterated,

M 2

though sometimes preserved, even in the white marble. Both the altered limestone and hardened slate contain garnets in many places, also ores of iron, lead, and copper, with some silver. These alterations occur equally, whether the granite invades the strata in a line parallel to the general strike of the fossiliferous beds, or in a line at right angles to their strike, as will be seen by the accompanying ground plan. *

The indurated and ribboned schists above mentioned, bear a strong resemblance to certain shales of the coal found at Russell's Hall, near Dudley, where coal mines have been on fire for ages. Beds of shale of considerable thickness, lying over the burning coal, have been baked and hardened so as to acquire a flinty fracture, the layers being alternately green and brick-coloured.

The granite of Cornwall, in like manner, sends forth veins into a coarse argillaceous-schist, provincially termed killas. This killas is converted into hornblende-schist near the contact with the veins. These appearances are well seen at the junction of the granite and killas, in St. Michael's Mount, a small island nearly 300 feet high, situated in the bay, at a distance of about three miles from Penzance.

The granite of Dartmoor, in Devonshire, says

* Keilhau, Gæa Norvegica, pp. 61—63.

Mr. De la Beche, has intruded itself into the slate and slaty sandstone called greywacké, twisting and contorting the strata, and sending veins into them. Hence some of the slate rocks have become " micaceous, others more indurated, and with the characters of mica-slate and gneiss, while others again appear converted into a hard-zoned rock strongly impregnated with felspar." *

We learn from the investigations of M. Dufrénoy, that in the eastern Pyrenees there are mountain masses of granite posterior in date to the formation called lias and chalk of that district, and that these fossiliferous rocks are greatly altered in texture, and often charged with iron-ore, in the neighbourhood of the granite. Thus in the environs of St. Martin, near St. Paul de Fénouillet, the chalky limestone becomes more crystalline and saccharoid as it approaches the granite, and loses all traces of the fossils which it previously contained in abundance. At some points also it becomes dolomitic, and filled with small veins of carbonate of iron, and spots of red iron-ore. At Rancié the lias nearest the granite is not only filled with iron-ore, but charged with pyrites, tremolite, garnet, and a new mineral somewhat allied to felspar, called, from the place in the Pyrenees where it occurs " couzeranite."

* Geol. Manual, p. 479.

M 3

Now the alterations above described as super-induced in rocks by volcanic dikes and granite veins, prove incontestably that powers exist in nature capable of transforming fossiliferous into crystalline strata, — powers capable of generating in them a new mineral character, similar, nay, often absolutely identical with that of gneiss, mica-schist, and other stratified members of the hypogene series. The precise nature of these altering causes, which may provisionally be termed plutonic, is in a great degree obscure and doubtful; but their reality is no less clear, and we must suppose the influence of heat to be in some way connected with the transmutation, if, for reasons before explained, we concede the igneous origin of granite.

The experiments of Gregory Watt, in fusing rocks in the laboratory, and allowing them to consolidate by slow cooling, prove distinctly that a rock need not be perfectly melted in order that a re-arrangement of its component particles should take place, and a partial crystallization ensue.* We may easily suppose, therefore, that all traces of shells and other organic remains may be destroyed; and that new chemical combinations may arise, without the mass being so fused as that the lines of stratification should be wholly obliterated.

* Phil. Trans. 1804.

We must not, however, imagine that heat alone, such as may be applied to a stone in the open air, can constitute all that is comprised in plutonic action. We know that volcanos in eruption not only emit fluid lava, but give off steam and other heated gases, which rush out in enormous volume, for days, weeks, or years continuously, and are even disengaged from lava during its consolidation. When the materials of granite, therefore, came in contact with the fossiliferous stratum in the bowels of the earth under great pressure, the contained gases might be unable to escape; yet when brought into contact with rocks, might pass through their pores with greater facility than water is known to do. (see p. 74.) These aeriform fluids, such as sulphuretted hydrogen, muriatic acid, and carbonic acid, issue in many places from rents in rocks, which they have discoloured and corroded, softening some and hardening others. If the rocks are charged with water, they would pass through more readily; for, according to the experiments of Henry, water, under an hydrostatic pressure of ninety-six feet, will absorb three times as much carbonic acid gas as it can under the ordinary pressure of the atmosphere. Although this increased power of absorption would be diminished, in consequence of the higher temperature found to exist as we descend in the earth, yet Professor Bischoff has shown that the heat by no means augments in such a proportion as to counteract the effect of augmented pressure. * There are other gases, as well as the carbonic acid, which water absorbs, and more rapidly in proportion to the amount of pressure. Now even the most compact rocks may be regarded, before they have been exposed to the air and dried, in the light of sponges filled with water; and it is conceivable that heated gases brought into contact with them, at great depths, may be absorbed readily, and transfused through their pores. Although the gaseous matter first absorbed would soon be condensed, and part with its heat, yet the continued arrival of fresh supplies from below, might, in the course of ages, cause the temperature of the water, and with it that of the containing rock, to be materially raised.

M. Fournet, in his description of the metalliferous gneiss near Clermont, in Auvergne, states that all the minute fissures of the rock are quite saturated with free carbonic acid gas, which rises plentifully from the soil there and in many parts of the surrounding country. The various elements of the gneiss, with the exception of the quartz, are all softened; and new combinations of the acid, with lime, iron, and manganese, are continually in progress. †

Another illustration of the power of subterra-

* Poggendorf's Annalen, No. XVI. Second Series, vol. iii.
† See Principles, *Index*, " Auvergne," &c.

nean gases is afforded by the stufas of St. Calo-
gero, situated in the largest of the Lipari Islands.
Here, according to the description lately published
by Hoffmann, horizontal strata of tuff, extending
for four miles along the coast, and forming cliffs
more than 200 feet high, have been discoloured
in various places, and strangely altered by the
" all-penetrating vapours."   Dark clays have become yellow, or often snow-white; or have assumed
a chequered and brecciated appearance, being
crossed with ferruginous red stripes.   In some
places the fumeroles have been found by analysis
to consist partly of sublimations of oxide of iron;
but it also appears that veins of calcedony and
opal, and others of fibrous gypsum, have resulted
from these volcanic exhalations. *

The reader may also refer to M. Virlet's account of the corrosion of hard, flinty, and jaspideous rocks near Corinth, by the prolonged agency
of subterranean gases †; and to Dr. Daubeny's description of the decomposition of trachytic rocks
in the Solfatara, near Naples, by sulphuretted
hydrogen and muriatic acid gases. ‡

Although in all these instances we can only
study the phenomena as exhibited at the surface,

* Hoffmann's Liparischen Inseln, p. 38. Leipzig, 1832.
† See Princ. of Geol.; and Bulletin de la Soc. Géol. de
France, tom. ii. p. 330.
‡ See Princ. of Geol.; and Daubeny's Volcanos, p. 167.

it is clear that the gaseous fluids must have made their way through the whole thickness of porous or fissured rocks, which intervene between the subterranean reservoirs of gas and the external air. The extent, therefore, of the earth's crust, which the vapours have permeated and are now permeating, may be thousands of fathoms in thickness, and their heating and modifying influence may be spread throughout the whole of this solid mass.

The above observations are calculated to meet some of the objections which have been urged against the metamorphic theory on the ground of the small power of rocks to conduct heat; for it is well known that rocks, when dry and in the air, differ remarkably from metals in this respect. It has been asked how the changes which extend merely for a few feet from the contact of a dike could have penetrated through mountain masses of crystalline strata several miles in thickness. Now it has been stated that the plutonic influence of the syenite of Norway, has sometimes altered fossiliferous strata for a distance of a quarter of a mile, both in the direction of their dip and of their strike. (See Fig. 126. p. 243.) This is undoubtedly an extreme case; but is it not far more philosophical to suppose that this influence may, under favourable circumstances, affect denser masses, than to invent an entirely new cause to account for effects merely differing in quantity, and not in kind? The metamorphic theory does not require us to affirm that some contiguous mass of granite has been the altering power; but merely that an action, existing in the interior of the earth at an unknown depth, whether thermal, electrical, or other, analogous to that exerted near intruding masses of granite, has, in the course of vast and indefinite periods, and when rising perhaps from a large heated surface, reduced strata thousands of yards thick to a state of semi-fusion, so that on cooling they have become crystalline, like gneiss. Granite may have been another result of the same action in a higher state of intensity, by which a thorough fusion has been produced; and in this manner the passage from granite into gneiss may be explained.

Some geologists are of opinion, that the alternate layers of mica and quartz, or mica and felspar, or lime and felspar, are so much more distinct in certain metamorphic rocks, than the ingredients composing alternate layers in many sedimentary deposits, that the similar particles must be supposed to have exerted a molecular attraction for each other, and to have thus congregated together in layers, more distinct in mineral composition than before they were crystallized.

In considering, then, the various data already enumerated, the forms of stratification in metamorphic rocks, their passage on the one hand into the fossiliferous, and on the other into the plutonic formations, and the conversions which can be ascertained to have occurred in the vicinity of granite, we may conclude that gneiss and mica-schist may be nothing more than altered mica-ceous and argillaceous sandstones, that granular quartz may have been derived from siliceous sand-stone, and compact quartz from the same mate-rials. Clay-slate may be altered shale, and granular marble may have originated in the form of ordi-nary limestone, replete with shells and corals, which have since been obliterated; and, lastly, calcareous sands and marls may have been changed into im-pure crystalline limestones.

" Hornblende-schist," says Dr. MacCulloch, " may at first have been mere clay; for clay or shale is found altered by trap into Lydian stone, a substance differing from hornblende-schist almost solely in compactness and uniformity of texture." * " In Shetland," remarks the same author, " argil-laceous-schist (or clay-slate), when in contact with granite, is sometimes converted into hornblende-schist, the schist becoming first siliceous, and ulti-mately, at the contact, hornblende-schist." †

The anthracite found associated with hypogene rocks may have been coal; for we know that, in the vicinity of some trap dikes, coal is converted into anthracite.

* Syst. of Geol., vol. i. p. 210.   † Ibid., p. 211.

The total absence of any trace of fossils has inclined many geologists to attribute the origin of crystalline strata to a period antecedent to the existence of organic beings. Admitting, they say, the obliteration, in some cases, of fossils by plutonic action, we might still expect that traces of them would oftener occur in certain ancient systems of slate, in which, as in Cumberland, some conglomerates occur. But in urging this argument, it seems to have been forgotten, that there are stratified formations of enormous thickness, and of various ages, and some of them very modern, all formed after the earth had become the abode of living creatures, which are nevertheless in certain districts entirely destitute of all vestiges of organic bodies. In some, the traces of fossils may have been effaced by water and acids, at many successive periods; and it is clear, that the older the stratum, the greater is the chance of its being non-fossiliferous, even if it has escaped all metamorphic action.

It has been also objected to the metamorphic theory, that the chemical composition of the secondary strata differs essentially from that of the crystalline schists, into which they are supposed to be convertible. * The " primary " schists, it is said, usually contain a considerable proportion of

---

* Dr. Boase, Primary Geology, p. 319.

potash or of soda, which the secondary clays, shales,
and slates do not, these last being the result of the
decomposition of felspathic rocks, from which the
alkaline matter has been abstracted during the
process of decomposition. But this reasoning pro-
ceeds on insufficient and apparently mistaken data;
for a large portion of what is usually called clay,
marl, shale, and slate does actually contain a certain and often a considerable proportion of alkali;
so that it is difficult in many countries to obtain
clay or shale sufficiently free from alkaline ingredients to allow of their being burnt into bricks or
used for pottery.

Thus the argillaceous shales, as they are called,
and slates of the old red sandstone, in Forfarshire
and other parts of Scotland, are so much charged
with alkali, derived from triturated felspar, that,
instead of hardening when exposed to fire, they
melt readily into a glass. They contain no lime,
but appear to consist of extremely minute grains
of the various ingredients of granite, which are
distinctly visible in the coarser-grained varieties,
and in almost all the interposed sandstones. These
laminated clays, marls, and shales might certainly,
if crystallized, resemble in composition many of
the primary strata.

There is also potash in the vegetable remains
included in strata, and soda in the salts by which

they are sometimes so largely impregnated, as in Patagonia.

Another objection has been derived from the alternation of highly crystalline strata with others having a less crystalline texture. The heat, it is said, in its ascent from below, must have traversed the less altered schists before it reached a higher and more crystalline bed. In answer to this, it may be observed, that if a number of strata differing greatly in composition from each other be subjected to equal quantities of heat, there is every probability that some will be more fusible than others. Some, for example, will contain soda, potash, lime, or some other ingredient capable of acting as a flux; while others may be destitute of the same elements, and so refractory as to be very slightly affected by a degree of heat capable of reducing others to semi-fusion. Nor should it be forgotten that, as a general rule, the less crystalline rocks do really occur in the upper, and the more crystalline in the lower part of each metamorphic series.

But it will be impossible for the reader duly to appreciate the propriety of the term metamorphic, as applied to the strata hitherto called primary, until I have shown in the second part of this work, that these crystalline strata have been formed at a great variety of distinct periods.

PART II.

CHAPTER XII.

ON THE DIFFERENT AGES OF THE FOUR GREAT CLASSES
OF ROCKS.

Aqueous, plutonic, volcanic, and metamorphic rocks, considered chronologically—Lehman's division into primitive
and secondary—Werner's addition of a transition class—
Neptunian theory—Hutton on igneous origin of granite—
How the name of primary was still retained for granite—
The term "transition," why faulty—The adherence to
the old chronological nomenclature retarded the progress
of geology—New hypothesis invented to reconcile the
igneous origin of granite to the notion of its high antiquity—Explanation of the chronological nomenclature
adopted in this work, so far as regards primary, secondary,
and tertiary periods.

IN the first part of this work the four great classes
of rocks, the aqueous, the volcanic, the plutonic
and the metamorphic, have been considered with
reference to their external characters, their mineral composition, and mode of origin ; and it now
remains to treat of the same classes with reference
to the different periods at which they were formed.
In speaking of the aqueous rocks, for example, it
has been shown that they are stratified, that some
are calcareous, others argillaceous, some made up

of sand, others of pebbles; that some contain fresh-water, others marine fossils, and so forth; but the student has still to learn which rocks, exhibiting some or all of these characters, have originated at one period of the earth's history, and which at another.

So in regard to the volcanic and plutonic form-ations, we have hitherto examined their mineral peculiarities, forms, and mode of origin, but have still to inquire into their chronological history.

Lastly, a more curious question will demand our attention, when we endeavour to ascertain the relative ages of the metamorphic rocks, the chro-nology of which may be said to be twofold, each formation having been deposited at one period, and having assumed a crystalline texture at an-other.

It was for many years a received opinion, that the formation of whole classes of rocks, such as the plutonic and metamorphic, began and ended before any members of the aqueous and volcanic orders were produced; and although this idea has long been modified, and is nearly exploded, it will be necessary to give some account of the an-cient doctrine, in order that beginners may under-stand whence part of the nomenclature of geology still partially in use was derived.

About the middle of the last century, Lehman, a German miner, proposed to divide rocks into three classes, the first and oldest to be called primitive, comprising the plutonic and metamorphic rocks; the next to be termed secondary, comprehending the aqueous or fossiliferous strata; and the remainder or third class, the supposed effect of " local floods, and the deluge of Noah," corresponding to our alluvium, ancient and modern. In the primitive class, he said, such as granite and gneiss, there are no organic remains, nor any signs of materials derived from the ruins of pre-existing rocks. Their origin, therefore, may have been purely chemical, antecedent to the creation of living beings, and probably coeval with the birth of the world itself. The secondary formations, on the contrary, which often contain sand, pebbles, and organic remains, must have been mechanical deposits, produced after the planet had become the habitation of animals and plants. This bold generalization, although anticipated in some measure by Steno, a century before, in Italy, formed at the time an important step in the progress of geology, and sketched out correctly some of the leading divisions into which rocks may be separated. About half a century later, Werner, so justly celebrated for his improved methods of discriminating the mineralogical characters of rocks, attempted to improve Lehman's classification, and with this view intercalated a class, called by him " the transition formations," between the primitive and secondary. Between these last he had discovered, in northern Germany, a series of strata, which in their mineral peculiarities were of an intermediate character, partaking in some degree of the crystalline nature of micaceous and clay-slate, and yet exhibiting here and there signs of a mechanical origin and organic remains. For this group, therefore, forming a passage between Lehman's primitive and secondary rocks, the name of transition was proposed. They consisted principally of clay-slate and an argillaceous sandstone, called greywacké, and partly of calcareous beds. It happened in the district which Werner first investigated, that both the primitive and transition strata were highly inclined, while the beds of the newer and fossiliferous rocks were horizontal. To these latter, therefore, he gave the name of *flötz*, or flat; and every deposit more modern than the chalk, or uppermost of the flötz series, was designated " the overflowed land," an expression which may be regarded as equivalent to alluvium. As the followers of Werner soon discovered that the inclined position of the " transition beds," and the horizontality of the flötz, or newer fossiliferous strata, were mere local accidents, they soon abandoned the term flötz; and the four divisions of the Wernerian school were then named primitive, transition, secondary, and alluvium.

As to the trappean rocks, although their igneous
origin had been already demonstrated by Arduino,
Fortis, Faujas, and others, and especially by Des-
marest, they were all regarded by Werner as
aqueous, and as mere subordinate members of the
secondary formations. *

This theory of Werner's was called the " Neptunian," and for many years enjoyed much popu-
larity. It assumed that the globe had been at
first invested by an universal chaotic ocean, holding the materials of all rocks in solution.   From
the waters of this ocean, granite, gneiss, and other
crystalline formations, were first precipitated;
and afterwards, when the waters were purged of
these ingredients, and more nearly resembled
those of our actual seas, the transition strata were
deposited.   These were of a mixed character, not
purely chemical, because the waves and currents
had already begun to wear down solid land, and
to give rise to pebbles, sand, and mud; nor entirely without fossils, because a few of the first
marine animals had begun to exist. After this
period, the secondary formations were accumulated
in waters resembling those of the present ocean,
except at certain intervals, when, from causes
wholly unexplained, a partial recurrence of the
" chaotic fluid" took place, during which various

* See Principles, vol. i. chap. iv.

trap rocks, some highly crystalline, were formed.
This arbitrary hypothesis rejected all intervention
of igneous agency, volcanos being regarded as
partial and superficial accidents, of trifling account
among the great causes which have modified the
external structure of the globe.

Meanwhile Hutton, a contemporary of Werner,
began to teach, in Scotland, that granite as well
as trap was of igneous origin, and had at various
periods intruded itself in a fluid state into different parts of the earth's crust. He recognized
and faithfully described many of the phenomena
of granitic veins, and the alterations produced by
them on the invaded strata, which have been
treated of in the ninth chapter. He, moreover,
advanced the opinion, that the crystalline strata
called primitive had not been precipitated from a
primæval ocean, but were sedimentary strata altered by heat. In his writings, therefore, and in
those of his illustrator, Playfair, we find the germ
of that metamorphic theory which has been already
expounded. *

At length, after much controversy, the doctrine
of the igneous origin of trap and granite made
their way into general favour; but although it
was, in consequence, admitted that both granite
and trap had been produced at many successive

* See chapters X. and XI.

periods, the term primitive or primary still continued to be applied to the crystalline formations
in general, whether stratified, like gneiss, or unstratified, like granite. The pupil was told that
granite was a primary rock, but that some granites were newer than certain secondary formations; and in conformity with the spirit of the
ancient language, to which the teacher was still
determined to adhere, a desire was naturally en-
gendered of extenuating the importance of those
more modern granites which new observations
were continually bringing to light.

A no less decided inclination was shown to
persist in the use of the term " transition," after
it had been proved to be almost as faulty in its
original application as that of flötz. The name
of transition, as already stated, was first given by
Werner, to designate a mineral character, intermediate between the metamorphic state and that
of an ordinary fossiliferous rock. But the term
acquired also from the first a chronological import, because it had been appropriated to sedimentary formations, which, in the Hartz and
other parts of Germany, were more ancient than
the oldest of the secondary series, and were characterized by peculiar fossil zoophytes and shells.
When, therefore, geologists found in other districts
stratified rocks occupying the same position, and
inclosing similar fossils, they gave to them also

the name of *transition*, according to rules which will be explained in the next chapter; yet, in many cases, such rocks were found not to exhibit the same mineral texture which Werner had called transition. On the contrary, many of them were not more crystalline than different members of the secondary class; while, on the other hand, these last were sometimes found to assume a semi-crystalline and almost metamorphic aspect, and thus, on lithological grounds, to deserve equally the name of transition. So remarkably was this the case in the Swiss Alps, that certain rocks, which had for years been regarded by some of the most skilful disciples of Werner to be transition, were at last acknowledged, when their relative position and fossils were better understood, to belong to the newest of the secondary groups! If under such circumstances the name of transition was retained, it is clear that it ought to have been applied without reference to the age of strata, and simply as expressive of a mineral peculiarity. The continued appropriation of the term to formations of a given date, induced geologists to go on believing that the ancient strata so designated bore a less resemblance to the secondary than is really the case, and to imagine that these last never pass, as they frequently do, into metamorphic rocks.

The poet Waller, when lamenting over the antiquated style of Chaucer, complains that —

> We write in sand, our language grows,
> And, like the tide, our work o'erflows;

But the reverse is true in geology; for here it is our work which continually outgrows the language. The tide of observation advances with such speed, that improvements in theory outrun the changes of nomenclature; and the attempt to inculcate new truths by words invented to express a different or opposite opinion, tends constantly, by the force of association, to perpetuate error; so that dogmas renounced by the reason still retain a strong hold upon the imagination.

In order to reconcile the old chronological views with the new doctrine of the igneous origin of granite, the following hypothesis was substituted for that of the Neptunists. Instead of beginning with an aqueous menstruum or chaotic fluid, the materials of the present crust of the earth were supposed to have been at first in a state of igneous fusion, until part of the heat having been diffused into surrounding space, the surface of the fluid consolidated, and formed a crust of granite. This covering of crystalline stone, which afterwards grew thicker and thicker as it cooled, was so hot, at first, that no water could exist upon it; but as the refrigeration proceeded, the aqueous vapour in the atmosphere was condensed, and, falling in rain, gave rise to the first *thermal ocean*. So high was the temperature of this boiling sea, that no aquatic beings could inhabit its waters, and its deposits were not only devoid of fossils, but, like those of some hot springs, were highly crystalline. Hence the origin of the primary or crystalline strata.

Afterwards, when the granitic crust had been partially broken up, land and mountains began to rise above the waters, and rains and torrents ground down rock, so that sediment was spread over the bottom of the seas. Yet the heat still remaining in the solid supporting substances was sufficient to increase the chemical action exerted by the water, although not so intense as to prevent the introduction and increase of some living beings. During this state of things some of the residuary mineral ingredients of the primæval ocean were precipitated, and formed deposits (the transition strata of Werner), half chemical and half mechanical, and containing a few fossils.

By this new theory, which was in part a revival of the doctrine of Leibnitz, published in 1680, on the igneous origin of the planet, the old ideas respecting the priority of all crystalline rocks to the creation of organic beings, were still preserved; and the notion, that all the semi-crystalline and partially fossiliferous rocks belonged to one period, while all the earthy and uncrystalline formations originated at a subsequent epoch, was also perpetuated.

N

It may or may not be true, as the great Liebnitz imagined, that the whole planet was once in a state of liquefaction by heat; but there are certainly no geological proofs that the granite which constitutes the foundation of so much of the earth's crust was ever in a state of universal fusion. On the contrary, all our evidence tends to show that the formation of granite, like the deposition of the stratified rocks, has been successive, and that different portions of granite have been in a melted state at distinct and often distant periods. One mass was solid, and had been fractured, before another body of granitic matter was injected into it, or through it, in the form of veins. In short, the universal fluidity of the crystalline foundations of the earth's crust, can only be understood in the same sense as the universality of the ancient ocean. All the land has been under water, but not all at one time; so all the subterranean unstratified rocks to which man can obtain access have been melted, but not simultaneously.

In the present work the four great classes of rocks, the aqueous, plutonic, volcanic, and metamorphic, will form four parallel, or nearly parallel, columns in one chronological table. They will be considered as four sets of monuments relating to four contemporaneous, or nearly contemporaneous, series of events. I have endeavoured, in the Frontispiece, to express the manner in which members of each of the four classes may have originated simultaneously at every geological period. According to this view, the earth's crust may have been continually remodelled, above and below, by aqueous and igneous causes, from times indefinitely remote. In the same manner as aqueous and fossiliferous strata are now formed in certain seas or lakes, while in other places volcanic rocks break out at the surface, and are connected with reservoirs of melted matter at vast depths in the bowels of the earth, — so, at every era of the past, fossiliferous deposits and superficial igneous rocks were in progress contemporaneously with others of subterranean and plutonic origin, and some sedimentary strata were exposed to heat and made to assume a crystalline or metamorphic structure.

It can by no means be taken for granted, that during all these changes the solid crust of the earth has been increasing in thickness. It has been shown, that so far as aqueous action is concerned, the gain by fresh deposits, and the loss by denudation, must at each period have been equal; and in like manner, in the inferior portion of the earth's crust, the acquisition of new crystalline rocks, at each successive era, may merely have counterbalanced the loss sustained by the melting of materials previously consolidated. As to the relative antiquity of the crystalline foundations of the earth's crust, when compared to the fossiliferous

and volcanic rocks which they support, I have already stated, in the first chapter, that to pronounce
an opinion on this matter is as difficult as at once
to decide which of the two, whether the foundations or superstructure of an ancient city built on
wooden piles may be the oldest. We have seen
that to answer this question, we must first be prepared to say whether the work of decay and restoration had gone on most rapidly above or below,
whether the average duration of the piles has exceeded that of the stone buildings, or the contrary.
So also in regard to the relative age of the superior and inferior portions of the earth's crust;
we cannot hazard even a conjecture on this point,
until we know whether, upon an average, the power
of water above, or that of fire below, is most effi-
cacious in giving new forms to solid matter.

After the observations which have now been
made, the reader will perceive that the term primary must either be entirely renounced, or, if retained, must be differently defined, and not made
to designate a set of crystalline rocks, some of
which may be newer than the secondary formations. In this work I shall follow most nearly
the method proposed by Mr. Boué, who has called
all fossiliferous rocks older than the secondary by
the name of primary, which thus becomes a substitute for the term transition, so far as regards
the aqueous strata. To prevent confusion, how-

ever, I shall always speak of these as the *primary fossiliferous* formations, because the word primary has hitherto been almost inseparably connected with the idea of a non-fossiliferous rock.

If we can prove any plutonic, volcanic, or metamorphic rocks to be older than the secondary formations, such rocks will also be primary, according to this system. Mr. Boué having with great propriety excluded the metamorphic rocks, *as a class*, from the primary formations, proposed to call them all " crystalline schists," restricting the name of primary to the older fossiliferous or transition strata.

As there are secondary fossiliferous strata, so we shall find that there are plutonic, volcanic, and metamorphic rocks of contemporaneous origin, which I shall also term secondary.

In the next chapter it will be shown that the strata above the chalk have been called tertiary. If, therefore, we discover any volcanic, plutonic, or metamorphic rocks, which have originated since the deposition of the chalk, these also will rank as tertiary formations.

It may perhaps be suggested that some metamorphic strata, and some granites, may be anterior in date to the oldest of the primary fossiliferous rocks. The opinion is certainly not improbable, and will be discussed in future chapters; but I may here observe, that when we arrange the four

classes of rocks in four parallel columns in one
table of chronology, it is by no means assumed
that these columns are all of equal length; one
may begin at an earlier period than the rest, and
another may come down to a later point of time.
In the small part of the globe hitherto examined,
it is hardly to be expected that we should have
discovered either the oldest or the newest of all
the four classes of rocks.   Thus, if there be primary, secondary, and tertiary rocks of the fossiliferous class, and in like manner primary, secondary, and tertiary plutonic formations, we may not
be yet acquainted with the most ancient of the
primary fossiliferous beds, or with the newest of
the plutonic, and so of the rest.

CHAPTER XIII.

In the last chapter I spoke generally of the chronological relations of the four great classes of rocks, and I shall now treat of the aqueous rocks in particular, or of the successive periods at which the different fossiliferous formations have been deposited.

Now there are three principal tests by which we determine the age of a given set of strata; first, superposition; secondly, mineral character; and, thirdly, organic remains. Some aid can occasionally be derived from a fourth kind of proof, namely, the fact of one deposit including in it fragments of a preexisting rock, which last may thus be shown, even in the absence of all other evidence, to be the older of the two.

Superposition. — The first and principal test of the age of one aqueous deposit, as compared to another, is relative position. It has been already stated, that where the strata are horizontal, the bed which lies uppermost is the newest of the whole, and that which lies at the bottom the most ancient. So, of a series of sedimentary formations, they are like volumes of history, in which each writer has recorded the annals of his own times, and then laid down the book, with the last written page uppermost, upon the volume in which the events of the era immediately preceding were comme-morated. In this manner a lofty pile of chronicles is at length accumulated; and they are so arranged as to indicate, by their position alone, the order in which the events recorded in them have occurred.

In regard to the crust of the earth, however, there are some regions where, as the student has already been informed, the beds have been dis-turbed, and sometimes reversed. (See pp. 113, 114.) But the experienced geologist will not be deceived by these exceptional cases. When he finds that the strata are fractured, curved, in-clined, or vertical, he knows that the original order of superposition must be doubtful, and he will endeavour to find sections in some neighbouring district where the strata are horizontal, or only slightly inclined. Here, it is impossible that they can have been extensively thrown over and turned upside down, for such a derangement cannot have taken place throughout a wide area without leaving manifest signs of displacement and dislocation.

Mineral character. — The same rocks may often be observed to retain for miles, or even hundreds of miles, the same mineral peculiarities, if we follow them in the direction of the planes of stratification. But this uniformity ceases almost immediately, if we pursue them in an opposite direction. In that case we can scarcely ever penetrate a stratified mass for a few hundred yards, much less several miles, without beholding a succession of extremely dissimilar calcareous, argillaceous, and siliceous rocks. These phenomena lead to the conclusion, that rivers and currents have dispersed the same sediment over wide areas at one period, but at successive periods have been charged, in the same region, with very different kinds of matter. The first observers were so astonished at the vast spaces over which they were able to follow the same homogeneous rocks in a horizontal direction, that they came hastily to the opinion, that the whole globe had been environed by a succession of distinct aqueous formations, disposed round the nucleus of the planet, like the concentric coats of an onion.   But although, in fact, some formations may be continuous over districts as large as half of Europe, or even more, yet most of them either terminate wholly within narrower limits, or soon change their lithological character.   Sometimes they thin out gradually, as if the supply of sediment had failed in that direction, or they come abruptly to an end, as if we had arrived at the borders of the ancient sea or lake which served as their receptacle.   It no less frequently happens that they vary in mineral aspect and composition, as we pursue them horizontally.   For example, we trace a limestone for a hundred miles, until it becomes more arenaceous, and finally passes into sand, or sandstone.   We may then follow this sandstone, already proved by its continuity to be of the same age, throughout another district a hundred miles or more in length.

Organic remains. — This character must be used as a criterion of the age of a formation, or of the contemporaneous origin of two deposits in distant places, under very much the same restrictions as the test of mineral composition.

First, the same fossils may be traced over wide regions, if we examine strata in the direction of their planes, although by no means for indefinite distances.   This might have been expected; for although many species of animals and plants have a wide geographical range, yet each species generally inhabits a small part only of the entire globe, and is often incapable of existing in other regions. But, in those cases where the fossils vary, the mineral character of the rock often remains constant; and, on the other hand, the fossils are sometimes uniform throughout spaces where the lithological nature of the rock is variable. In this manner we are frequently enabled to prove the contemporaneous origin of the same formation by one test, when the other fails.

Secondly, while the same fossils prevail in a particular set of strata for hundreds of miles in a horizontal direction, we seldom meet with the same remains for many fathoms, and scarcely ever for several hundred yards, in a vertical line, or a line transverse to the strata. This fact has now been verified in almost all parts of the globe, and has led to a conviction, that at successive periods of the past, the same area of land and water has been inhabited by species of animals and plants as distinct as those which now people the antipodes, or which now coexist in the arctic, temperate, and tropical zones. It appears, that from the remotest periods there has been ever a coming in of new organic forms, and an extinction of those which pre-existed on the earth; some species having endured for a longer, others for a shorter time; but none having ever reappeared after once dying out. The law which has governed the creation and extinction of species seems to be expressed in the verse of the poet,

Natura il fece e poi ruppe la stampa. — *Ariosto.*
Nature made it, and then broke the die.

And this circumstance it is, which confers on fossils their highest value as chronological tests, giving to each of them, in the eyes of the geologist, that authority which belongs to contemporary medals in history.

The same cannot be said of each peculiar variety of rock; for some of these, as red marl and red sandstone, for example, may occur at once at the top, bottom, and middle of the entire sedimentary series; exhibiting in each position so perfect an identity of mineral aspect as to be undistinguishable. Such exact repetitions, however, of the same mixtures of sediment have not often occurred, at distant periods, in precisely the same parts of the globe; and even where this has happened, we may usually avoid confounding together the monuments of remote eras, by the aid of fossils and relative position.

Test by included fragments of older rocks. — It was stated, that independent proof may sometimes be obtained of the relative date of two formations, by fragments of an older rock being included in a newer one. This evidence may sometimes be of great use, where a geologist is at a loss to determine the relative age of two formations, from want of clear sections exhibiting their true order of position, or because the strata of each group are vertical. In such cases we sometimes discover that the more modern rock has been in part derived from the degradation of the older. Thus, for example, we may find chalk with flints; and, in another part of the same country, a distinct series, consisting of alternations of clay, sand, and pebbles. If some of these pebbles consist of flints, with fossil shells of the same species as those in the chalk, we may confidently infer that the chalk is the oldest of the two formations.

The number of groups into which the fossili-ferous strata may be separated, are more or less numerous, according to the views of classification which different geologists entertain; but when we have adopted a certain system of arrangement, we immediately find that a few only of the entire series of groups occur one upon the other in any single section or district.

The thinning out of individual strata was before described (p. 37.). But let the annexed diagram

Fig. 127.

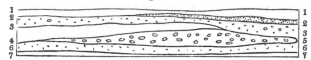

represent seven fossiliferous groups, instead of as
many strata. It will then be seen that in the middle
all the superimposed formations are present; but
in consequence of some of them thinning out, No. 2.
and No. 5. are absent at one extremity of the section, and No. 4. at the other.

If the reader consults the Frontispiece, he will
see, that as the strata A rest unconformably upon
the older groups, *a, b, c, e, f, g,* we should meet
with a very different succession in a vertical section exposed at different places; in one spot A
lying immediately on *c,* in another on *g,* and so
forth. Now here the difference has been partly
occasioned by denudation; the formations *a, b,* for
instance, once extended much farther to the
left, and but for denudation would have been
everywhere interposed between A and the rocks
e, f, g. In many instances the entire absence of
one or more formations of intervening periods between two groups, such as A and *c,* (see Frontispiece,) arises, not from the destruction of what
once existed, by denudation, but because no strata
of an intermediate age were ever deposited on *c.*
They were not formed at that place, either because the region was dry land during the interval,
or because it was part of a sea or lake to which
no sediment was carried.

In order, therefore, to establish a chronological

succession of fossiliferous groups, a geologist must begin with a single section, in which several sets of strata lie one upon the other. He must then trace these formations, by attention to their mineral character and fossils, continuously, as far as possible, from the starting point. As often as he meets with new groups, he must ascertain by superposition their age relatively to those first examined, and thus learn how to intercalate them in a tabular arrangement of the whole.

By this means the German, French, and English geologists have determined the succession of strata throughout a great part of Europe, and have adopted pretty generally the following groups, almost all of which have their representatives in the British Islands.

Groups of Fossiliferous Strata observed in Western Europe, arranged in what is termed a descending series, or beginning with the newest.

1. Newer Pliocene.		
2. Older Pliocene.	}	Tertiary or Supracretace-
3. Miocene.		ous.*
4. Eocene.		

---

* For tertiary, Mr. De la Beche has used the term " supracretaceous," a name implying that the strata so called are superior in position to the chalk.

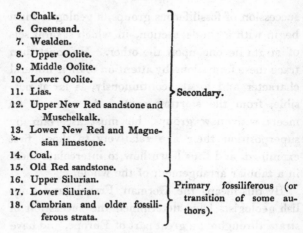

5. Chalk.
6. Greensand.
7. Wealden.
8. Upper Oolite.
9. Middle Oolite.
10. Lower Oolite.
11. Lias. } Secondary.
12. Upper New Red sandstone and Muschelkalk.
13. Lower New Red and Magnesian limestone.
14. Coal.
15. Old Red sandstone.
16. Upper Silurian.
17. Lower Silurian. } Primary fossiliferous (or transition of some authors).
18. Cambrian and older fossiliferous strata.

A glance at the above table will show that the three great sections called primary fossiliferous, secondary, and tertiary, are by no means of equivalent importance, if the eighteen subordinate groups comprise monuments relating to equal portions of past time, or of the earth's history. But this we cannot assert; but merely know that they each relate to successive periods, during which certain animals and plants, for the most part peculiar to that era, flourished, and during which different kinds of sediment were deposited in the space now occupied by Europe.

If we were disposed, on palæontological grounds, to divide the entire fossiliferous series into a few groups, less numerous than those in the above table, and more nearly co-ordinate in value than the sections called primary, secondary, and tertiary, we might, perhaps, adopt the six following groups or periods. *   At the same time I may observe, that in the present state of the science, when we have not yet compared the evidence derivable from all classes of fossils, not even those most generally distributed, such as shells, corals, and fish, such generalizations are premature, and can only be regarded as conjectural schemes for the founding of large natural groups.

1. Tertiary - - -	from the Newer Pliocene to the Eocene inclusive.
2. Cretaceous - -	from the Chalk to the Wealden inclusive.
3. Oolitic - - -	from the Oolite to the Lias inclusive.
4. Upper New Red - -	including the Keuper, Muschelkalk, and Bunter Sandstein of the Germans.
5. Lower New Red and Carboniferous - -	including Magnesian Limestone (Zechstein), Coal, and Old Red sandstone.
6. Primary fossiliferous -	from the Upper Silurian to the oldest fossiliferous rocks inclusive.

The limits of this volume will not allow of a full description, even of the leading features of all the formations enumerated in the above tables; but I shall briefly advert to each of them in chro-

* Palæontology is the science which treats of fossil remains, both animal and vegetable.   *Etym.* παλαιος, *palaios*, ancient, οντα, *onta*, beings, and λογος, *logos*, a discourse.

nological order, as they will afford illustrations of
the rules of classification, the tests of relative age,
and the mode of deriving information from geological monuments respecting the former history of
the earth and its inhabitants.

Tertiary formations. — These strata, as we have
seen, were so called because, when first discovered,
they were observed to be of a date posterior to
the chalk, which had long been regarded as the
last or uppermost of the secondary formations.
It was remarked, that in France, Italy, Germany,
and England, the tertiary deposits occupied a position, in reference to all older rocks, like that of the
waters of lakes, inland seas, and gulfs in relation
to a continent, being often, like such waters, of
great depth, though of limited area, and frequently
occurring in detached and isolated patches. The
strata were for the most part horizontal, but
usually surrounded by older rocks, of which the
beds were highly inclined or vertical.

On comparing together the fossils of the aqueous
formations in general, especially the testacea, which
are the most abundant and best preserved of all,
it appears that those of the primary fossiliferous
rocks depart most widely in form and structure
from the type of the living creation, those of the
secondary less widely, and the tertiary least of all.
In like manner, if we divide the tertiary deposits
into four principal groups, and then compare the

fossil shells which they contain with the testacea now living in the nearest seas in the same latitudes, we find that the shells of the oldest strata have much less resemblance, on the whole, to the fauna of the neighbouring seas, than those of the newest group. In a word, in proportion as the age of a tertiary formation is more modern, so also is the resemblance greater of its fossil shells to the testaceous fauna of the actual seas.

Having observed the prevalence of this change of character in the tertiary strata of France and Italy, in 1828, I conceived the idea of classing the whole series of tertiary strata into four groups, endeavouring to find characters for each expressive of their different degrees of affinity to the living fauna. I hoped that an estimate of this varying relation to the fauna of the existing seas might be obtained by determining the proportional number of shells identical with living species which belonged to each group. With this view, I obtained information respecting the specific identity of many tertiary and recent shells from several Italian naturalists; and among others, from Professors Bonelli, Guidotti, and Costa.

I have explained at length, in the Principles of Geology, the opinions which were at that time generally entertained respecting the classification of tertiary formations, and the observations which led me, in 1828, to divide them into four groups, by reference not only to their geological position, but also to the proportional number of recent species found fossil in each. I have also there stated, that having, in 1829, become acquainted with M. Deshayes, of Paris, I learnt from him that he had arrived, by independent researches, and by the study of a large collection of fossil and recent shells, at very similar views. At my request he drew up, in a tabular form, lists of all the shells known to him to occur both in some tertiary formation and in a living state, for the express purpose of ascertaining the proportional number of fossil species identical with the recent which characterized the successive groups; and this table was published by me in 1833. * The number of tertiary fossil shells examined by M. Deshayes was about 3000; and the recent species with which they had been compared, about 5000. The result at which that naturalist arrived was, that in the oldest tertiary deposits, such as those found near London and Paris, there were about $3\frac{1}{2}$ per cent. of species of fossil shells identical with recent species; in the next, or middle tertiary period, to which certain strata on the Loire and Gironde, in France, belonged, about 17 per cent.; and in the deposits of a third, or newer era, embracing those of the Subapennine hills, from 35 to 50 per cent.

---

* See Princ. of Geol. vol. iii., 1st ed.

In formations still more modern, some of which I had particularly studied in Sicily, where they attain a vast thickness and elevation above the sea, the number of species identical with those now living was from 90 to 95 per cent. For the sake of clearness and brevity, I proposed to give short technical names to these four groups, or the periods to which they respectively belonged. I called the first or oldest of them Eocene, the second Miocene, the third Older Pliocene, and the last or fourth Newer Pliocene. The first of the above terms, Eocene, is derived from ηως, eos, *dawn*, and καινος, cainos, *recent*, because the fossil shells of this period contain an extremely small proportion of living species, which may be looked upon as indicating the dawn of the recent or existing state of the testaceous fauna.

The other terms, Miocene and Pliocene, are comparative; the first meaning less recent, (from μειον, meion, *less*, and καινος, cainos, *recent*,) and the other more recent, (from πλειον, pleion, *more*, and καινος, cainos, *recent*,) they express the *more* or *less* near approach which the deposits of these eras, when contrasted with each other, make to the existing creation, at least so far as the mollusca are concerned. It may assist the memory of students to remind them, that the *Mi*ocene contain a *mi*nor proportion, and *Pli*ocene a comparative *plu*rality of recent species; and that the greater num-

ber of recent species always implies the more
modern origin of the strata.

Two subjects of discussion have arisen respecting the tables above alluded to; first, whether the
fossil shells were, upon the whole, correctly identified with recent species by M. Deshayes; secondly,
whether such a per-centage of recent species occurring fossil in particular groups, affords the best
criterion for estimating the relation of each fossil
fauna to the living creation.

Now in regard to the per-centage test, its application must evidently depend on the extent to
which conchologists are agreed in their determin-
ation of species. In every branch of natural history there is always some difference of opinion as
to certain species which are variable in their
characters, and seem to pass by imperceptible
gradations into other forms, considered by many
zoologists and botanists as entitled to rank as distinct species. The difficulty of defining the limits
in such cases is not greater, perhaps, in conchology
than in other departments; but it happens that
this science has advanced very rapidly since the
year 1830, when M. Deshayes drew up the tables
published in the Principles of Geology. In that
year he had it in his power to refer to no more
than 5000 species of recent shells then in Paris;
but the number of species now in the public and
private collections of Europe has increased to

between 8000 and 9000; and, what is of no less consequence, individuals of species which before that time were extremely rare, have been supplied in abundance. Fossil shells also have been collected with equal zeal and success; and thus the facility of discriminating nice distinctions in closely allied species, or of deciding which characters are constant and which variable, has been greatly promoted; and the study of these more ample data has led all conchologists to separate many species, both of fossil and recent shells, which before they had confounded together.

In consequence of the changes of opinion brought about by these additions to our knowledge, it has become necessary not only to examine all the newly discovered fossil and recent testacea, but also to reconsider all the species previously known. As this laborious task has not yet been executed by M. Deshayes, engaged as he is in other scientific labours, I am unable at present to offer to the reader the improved results which the revision of the tables drawn up in 1830 would afford. In the mean time I have obtained the aid of several eminent conchologists, and in particular of Dr. Beck, of Copenhagen, in comparing a great number of the recent and fossil shells which had been identified. By this investigation I have come to the conclusion that the per-centage of recent species in a fossil state is decidedly less,

especially in the older tertiary strata, than was indicated in the list published in 1833. A large
number, in particular, of the forty-two species of
Eocene testacea, to which the names of recent
shells were given in the tables, cannot be considered as identical, if we adopt the same standard
of specific distinctions as is recognized in the new
edition of Lamarck's conchology, edited by M.
Deshayes himself, in 1836.

But although many corrections are indispensable, and the proportion of recent species found
fossil in the Eocene, Miocene, and older Pliocene
strata may be considerably less than was at first
supposed, we have no reason on this account to
feel discouraged in an attempt to found the classification and nomenclature of the tertiary periods
on the great principle before explained; namely,
the comparative resemblance of the testaceous
fauna of each period to that of the neighbouring
seas. There can be no cabalistic virtue in such
numbers as 3. 17. or 40., which were at first
imagined to express correctly the proportional
number of identical species in three of the tertiary
periods; but until the time arrives when we can
obtain the general acquiescence of conchologists
as to the real proportional numbers, we must endeavour to find some readier method of estimating the relation of one fauna to another; a method
not involving the question of the identity or non-

identity of every fossil with some known recent species.

Now, it has been suggested by Dr. Beck that, in order to form such an estimate of the comparative resemblance of the faunas of different eras, we may follow the same plan as would enable us to appreciate the amount of agreement or discrepancy between the faunas now existing in two distinct geographical regions.

It is well known that, although nearly all the species of mollusca inhabiting the temperate zones on each side of the equator are distinct, yet the whole assemblage of species in one of these zones bears a striking analogy to that in the other, and differs in a corresponding manner from the tropical and arctic faunas. By what language can the zoologist express such points of agreement or disagreement, where the species are admitted to be distinct?

In such cases it is necessary to mark the relative abundance in the two regions compared of certain families, genera, and sections of genera; the entire absence of some of these, the comparative strength of others, this strength being sometimes represented by the numbers of species, sometimes by the great abundance and size of the individuals of certain species. It is, moreover, important to estimate the total number of species inhabiting a given area; and also the average

o

proportion of species to genera, as this differs materially according to climate. Thus, if we adopt comprehensive genera like those of Lamarck, we shall find, according to Dr. Beck, that, upon an average, there are in arctic latitudes nearly as many genera as species; in the temperate regions, about three or four species to a genus; in the tropical, five or six species to a genus.

The method of which the above sketch conveys but a faint outline, is the more easy of application to the tertiary deposits of Europe, because the conchological fauna of the Eocene period indicates a tropical climate; that of the Miocene strata, a climate bordering on the tropics; and that of the Older and Newer Pliocene deposits, a climate much more closely approaching to, if not the same as, that of the seas in corresponding latitudes.

Although I cannot enter in this work into farther details, it may be stated that, if we compare tertiary formations on this principle, the nomenclature above proposed will not be inappropriate; for the fauna of the older, or Eocene, tertiary formations is still the first in the order of time in which there is an assemblage of testacea like that of the present ocean between the tropics; and in this period a small proportion of mollusca are undistinguishable from living species; whereas at the opposite extreme of the series, or in the Newer Pliocene deposits, all conchologists agree that the marine shells are all, or nearly all, identical with those now inhabiting the nearest seas. As to the Miocene and Older Pliocene groups, the terms less and more will always express correctly the different degrees of analogy which their fossils bear to the assemblage of living species in similar latitudes.

But it should never be forgotten that, as the extinct species preponderate in all groups, with the exception of the Newer Pliocene, it is from their characters that we derive the distinguishing feature in the palæontology of each period. The relative approach which the shells may make to the living fauna affords a useful and interesting term of comparison; but it is one feature only, and by no means the most prominent one, in the organic remains of successive periods.

CHAPTER XIV.

RECENT and Newer Pliocene strata. — If we begin
with the history of the more modern aqueous formations, and then pass on to the more ancient, the
first strata which present themselves are those
termed, in the last chapter, the Newer Pliocene.
But in what manner shall we define the limits between this group and those fossiliferous deposits
which are now in progress, or which have accumulated under water since the globe was inhabited by
man? The strata last mentioned, namely, those
of the human period, I shall call *Recent*, distinguishing them from the most modern tertiary
formations. Strata may be proved to belong to
the Recent period by our finding in them the

bones of man in a fossil state, that is to say, imbedded in them by natural causes; or we may
recognize them by their containing articles fabricated by the hands of man, or by showing that
such deposits did not exist in the place where we
now observe them at a given period of the past
when man existed, so that they must be of subsequent origin. In general all recent formations lie
hidden from our sight beneath the waters of lakes
and seas; but we may examine them wherever
these lakes or seas have been partially converted
into land, as in the deltas of rivers, or where the
submerged ground has been heaved up by subterranean movements, and laid dry.

Thus at Puzzuoli, near Naples, marine strata
are seen containing fragments of sculpture, pottery, and the remains of buildings, together with
innumerable shells retaining in part their colour,
and of the same species as those now inhabiting
the Mediterranean. The uppermost of these beds
is about twenty feet above the level of the sea.
Their emergence can be proved to have taken
place since the beginning of the sixteenth century.* But the hills at the base of which these
strata have been deposited, and those of the interior of the adjacent country round Naples, some
of which rise to the height of 1500 feet above the

* See Principles, Index, " Serapis."

sea, are formed of horizontal strata of the Newer
Pliocene period; that is to say, the marine shells
observed in them are of living species, and yet
are not accompanied by any remains of man or
his works. Had such been discovered, it would
have afforded to the antiquary and geologist matter of great surprise, since it would have shown
that man was an inhabitant of that part of the
globe, while the materials composing the present
hills and plains of Campania were still in the
progress of deposition at the bottom of the sea;
whereas we know that for nearly 3000 years, or
from the times of the earliest Greek colonists, no
extensive revolution in the physical geography of
that part of Italy has occurred.

In Sweden, analogous phenomena have been observed. Near Stockholm, for example, when the
canal of Södertelje was dug, horizontal beds of
sand, loam, and marl were passed through, in
some of which the same peculiar assemblage of
testacea which now live in the Baltic were found.
Mingled with these, at different depths, were detected various works of art implying a rude state
of civilization, and some vessels built before the
introduction of iron. These vessels and imple-
ments must have sunk to the bottom of an arm of
the sea, afterwards filled up with sand and loam
including marine shells, and the whole must then
have been upraised; so that the upper beds became

sixty feet higher than the surface of the Baltic.
There are, however, in the neighbourhood of these
formations, others precisely similar in mineral composition and testaceous remains, which ascend to
the height of between 100 and 200 feet, in which
no vestige of human art has been seen. Similar deposits reach an elevation of 500 and even 600 feet
in Norway, as in the neighbourhood of Christiania,
where they have usually been described as raised
beaches, but are, in fact, strata of clay, sand, and
marl, often many hundred feet thick, which cover
the inland country far and wide, filling valleys
and deep depressions in the granite, gneiss, and
primary fossiliferous rocks, just as the tertiary
formations of England and France rest upon the
chalk, or fill depressions in it.

All conchologists are agreed that the shells of
the deposits above mentioned are nearly all, perhaps all, absolutely identical with those now peo-
pling the contiguous ocean; so that, in the absence
of any evidence of their being Recent, we must
regard them as Newer Pliocene formations.

Along the western shores of South America,
Recent and Newer Pliocene strata have in like
manner been brought to light. These often consist of enormous masses of shells, similar to those
now swarming in the Pacific. In one bed of this
kind, in the island of San Lorenzo, near Lima,
Mr. Darwin found, at the altitude of eighty-five

feet above the sea, pieces of cotton-thread, plaited rush, and the head of a stalk of Indian corn, all of which had evidently been imbedded with the shells. At the same height on the neighbouring mainland, he found other signs corroborating the opinion that the ancient bed of the sea had there also been uplifted eighty-five feet, since the region was first peopled by the Peruvian race. * But similar shells, or strata containing them, have been found much higher, almost every where between the Andes and sea coasts of Chili and Peru, in which no human remains were ever, or in all probability ever will be, discovered. These strata, therefore, may provisionally, at least, be designated Newer Pliocene.

In the West Indies, also, rocks both of the Recent and Newer Pliocene periods abound. Thus, a solid limestone occurs at the level of the sea-beach in the island of Guadaloupe, enveloping human skeletons. The stone is extremely hard, and chiefly composed of comminuted shell and coral, with here and there some entire corals and shells, of species now living in the adjacent sea. With them are included arrow heads, fragments of pottery, and other fabricated articles. A limestone with similar contents has been formed, and is still forming, in St. Domingo and other islands.

* Journal, p. 451. (for title, see note, p. 137.)

But there are also more ancient rocks in the West Indian Archipelago, as in Cuba, near the Havanna, and in other islands in which are shells identical with those now living in corresponding latitudes; some well preserved, others in casts, all referable to a period which, if we can depend on negative evidence, was anterior to the introduction of man into the New World.

The history of Holland, during the last 2000 years, makes us acquainted with a vast accession of Recent strata, by which parts of the sea near the mouths of the Rhine have been filled up and converted into dry land. But, if we ascend the Rhine, we find throughout its course, from Cologne to the frontiers of Switzerland, a yellow calcareous loam, called loess by the Germans, in which are fossil shells, both freshwater and terrestrial, of common European species. The entire thickness of this loam amounts in some places to 200 or 300 feet, and it rises from the height of 300 to 1200 feet above the sea. Bones of the mammoth or extinct elephant, together with those of the horse, and some other quadrupeds, have been met with in this Newer Pliocene formation, but no remains or signs of man; and it can be proved that the physical geography of the whole valley of the Rhine has undergone enormous changes since the deposition of this loam.

No marine strata of the Recent period have yet

been brought to light in England which rise to such a height above the level of the sea as the highest tides may not once have reached. Buried ships have been found in the former channels of the Rother in Sussex, of the Mersey in Kent, and Thames near London. Canoes and stone-hatchets have been dug up, in almost all parts of the kingdom, in peat and shell-marl; but there is no evidence, as in Sweden, Italy, Peru, Chili, and other parts of the world, of the bed of the sea, and the adjoining coast, having been uplifted bodily in modern times, so that Recent formations have become land. There are, however, in various parts of Great Britain and Ireland, Newer Pliocene deposits of marine origin, consisting of sand and clay, usually of small thickness; as, for example, in Cornwall, and near the borders of the great estuaries of the Clyde and Forth, in Scotland, and in that of the Shannon, in Ireland. These are found usually near the coast, but in some rare instances they penetrate inland to a distance of sixty miles from the sea, as at Bridgnorth, in Shropshire.* They also rise occasionally to great heights, as at Preston, in Lancashire, where they are 350 feet above the sea; and, what is still more remarkable, on a mountain called Moel Tryfane, in Wales, near the Menai Straits, they attain an

* See Murchison, Proceedings of Geol. Soc., vol. ii. p. 333.

elevation of about 1400 feet. *   In all these places they contain shells indisputably of the same species as those which now people the British seas; and although, perhaps, on more accurate examination some slight intermixture of extinct testacea will appear, yet the geologist will always refer them to the most modern tertiary era.

There are, moreover, a great many freshwater deposits scattered over England, which belong to the Newer Pliocene period, as at North Cliff, in the county of York, where thirteen species of British land and freshwater shells were found imbedded in the same strata with the remains of the bison and mammoth.† In like manner, at Cropthorne, in Worcestershire, on the banks of the Avon, a tributary of the Severn, Mr. Strickland observed fluviatile and land shells, nearly all of recent species, with the bones of an extinct kind of hippopotamus. Recent freshwater shells also appear in beds of loam, together with bones of the deer and mammoth, in the cliffs of the estuary of the Stour, in Suffolk. Some writers have confounded these and similar fluviatile and lacustrine strata, with the ancient alluviums which they term diluvial.

Older Pliocene strata in England — Crag. — There are some few countries in Europe, as in

* Proceedings of Geol. Soc., vol. i. p. 331., and vol. ii. p. 333.
† See Principles, Index, " Mammoth."

the district between the Gironde and the Pyrenees, in the south of France, or that between the Alps, north of Vicenza, and the hills near Turin, in the north of Italy, where fossiliferous strata representing all the three periods, the Eocene, Miocene, and Older Pliocene, are present. But the tertiary deposits of England are limited to the Eocene and the Older and Newer Pliocene groups, the Miocene being wanting.

It is chiefly in the eastern part of the county of Suffolk that a deposit provincially named crag is seen in its most characteristic form. This crag consists chiefly of a series of thin layers of quartzose sand and comminuted shell, which rest sometimes on chalk, sometimes on an Eocene tertiary formation, called the "London Clay." Mr. Charlesworth, whose opinion I have lately had opportunities of confirming, has correctly stated that the crag, in part of Suffolk, may be divided into two distinct masses, the upper of which may be termed the red, and the lower the coralline crag.* The inferior division, however, is of very limited extent, ranging over an area about twenty miles in length, and three or four miles in breadth, between the rivers Alde and Stour.

The red crag is generally at once distinguishable from the coralline, by the deep red ferruginous

* London and Edin. Phil. Mag. No. 38. p. 81. Aug. 1835.

or ochreous-colour of its sands and fossils. Its strata are also remarkable for the oblique or diagonal position of the subordinate layers (see p. 38.); and these often consist of small flat pieces of shell, which lie parallel to the planes of the smaller strata, showing clearly that they were so deposited, and that this structure has not been due to any subsequent rearrangement of the mass after deposition. That the ancient sandbanks in question had sometimes sides sloping in all directions, is implied by the fact that the oblique layers sometimes slant towards all points of the compass in different parts of the same quarry. They were probably shifting sands, and a great proportion of the shells composing them have been ground down to small pieces, while others have been rolled; and the two parts of the bivalves are almost invariably disunited. The red crag contains some peculiar fossils, and others which seem to have been washed out of the lower or coralline crag. Some few of the bivalves of the red crag are entire, with both valves joined.

The coralline crag is usually free from ferruginous stains, and consists of light greenish shelly marl, and white calcareous sand. Sometimes it forms a soft building stone, in which entire shells, echini, and many zoophytes are imbedded. Here and there the softer mass is divided by thin flags of hard limestone, in which are corals in a good state

of preservation, which evidently grew at the bottom
of a tranquil sea, in the position in which we now
see them. Yet the sands in this formation, as in
the red crag, are often composed entirely of comminuted shell.

In some places, as at Tattingstone, in Suffolk,
the lithological distinction of the two divisions,
though perceptible, is much less marked; the inferior crag being composed chiefly of greenish
marl, with only a few stony beds. The shells also
are mostly broken, and corals are almost as rare
as in the red crag.

At some places, as near Orford, the coralline
crag is exposed at the surface, and the bottom of
it has not been reached at the depth of fifty feet.
Yet not far from this town, the surface is occupied
exclusively with red crag, which rests immediately
upon the London clay. Wherever the two divisions are found together, the coralline mass is the
lower of the two, and is interposed between the
red crag and London clay; and the strata of the
upper and lower crag are unconformable one to
the other, as in the section represented in the
annexed diagram, which I have myself examined.

Fig. 128.

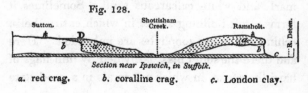

Section near Ipswich, in Suffolk.

a. red crag.   b. coralline crag.   c. London clay.

In the places here referred to, the coralline crag varies in thickness from fifteen to more than twenty feet, and the red crag is often much thicker.

Amongst upwards of 400 species of testacea found in the crag, there are many common to both divisions; but some, which are very abundant in the red, have never been met with in the coralline crag; as, for example, the *Fusus contrarius* (Fig. 129.), and several species of Nassa and

Fossils characteristic of the Red Crag.

Fig. 129.

Fig. 130.

Fig. 131.

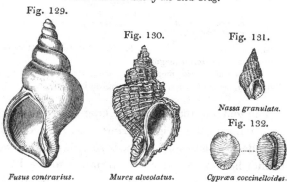

Nassa granulata.

Fig. 132.

Fusus contrarius.   *Murex alveolatus.*   *Cypræa coccinelloides.*

Murex (see Figs. 130, 131.), which two genera seem never to have been discovered in the lower crag. On the other hand, scarcely any corals have been found in the red crag. These abound in the inferior division, and some of them are of a globular form, and belong to the genera Theonoa (Lamoroux), Cellepora, and a third to Fascicularia, which

is of very peculiar structure, unknown in the living creation.   (See Fig. 133.)

Fig. 133.

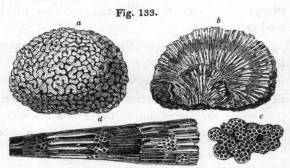

Fascicularia aurantium, Milne Edwards.   Family, *Tubuliporidæ*,
of same author.   *A coral of an extinct genus, from the inferior
or coralline crag, Suffolk.*

a. exterior.   *b.* vertical section of interior.

c. portion of exterior magnified.

d. portion of interior magnified, showing that it is made up of
long, thin, straight tubes, united in conical bundles.

The general analogy of the crag shells to those
now living in the neighbouring seas, between the
latitudes 50° and 60° north, is so striking that we
cannot hesitate to refer the formation to the Pliocene period; but, as all conchologists are agreed that
more than half the species are extinct or unknown,
it is to the Older and not the Newer Pliocene
period that they belong.   Dr. Beck, after examining 260 species of these shells, informs me that
the average number of species to genera is such
as indicates a temperate climate, a result which is
also confirmed by the large development of cer-

tain northern forms, such as the genus Astarte
(see Fig. 134.), of which there are about fourteen
species, many of them being rich in individuals;
and there is an absence of genera peculiar to hot
climates, such as Conus, Oliva, Mitra, Fasciolaria,

Fig. 134.

Astarte, (*Crassina*, Lam.) ; *species common to upper and lower crag.*

and others.   The cowries (*Cypræa*) (Fig. 132.)
also are small, as in the colder regions.   A large
volute, called *Voluta Lamberti* (Fig. 135.), may
Fig. 135.   seem an exception ; but it differs in
form from the volutes of the torrid
zone, and may, like the large *Voluta
Magellanica*, have been extra-tropical.

When I first submitted the shells of
the crag to M. Deshayes, in 1829, he
recognized their general resemblance
to the fauna of the German ocean, and
Voluta Lamberti,   determined that out of 111 species
young individ.   there were 45 identical with those
now living.   Dr. Beck, on the other hand, who
has since seen much larger collections, considers
that almost all the species are distinguishable from

those now living, and this subject is still under discussion.

It has been asked whether, as the upper and lower crag of Suffolk differs greatly in mineral composition and fossils, they may not belong to two different tertiary periods. To this I may reply, that the general character of the shells is the same, and by no means leads to such a conclusion. The two deposits may have been going on contemporaneously under different geographical conditions in the same sea. One region of deep and clear water, far from the shore, may have been fitted for the growth of certain corals, echini, and testacea; while another shallower part nearer the shore, and more frequently turbid, or where sand and shingle were occasionally drifted along, may have been favourable to other species. After this, the region of deep and tranquil water becoming shallow, or exposed to the action of waves and currents, a formation like the coralline crag may have been covered over with sandy deposits, such as the red crag, and many fossils of the older beds may have been washed into the newer strata. If a considerable lapse of time intervened in a particular spot between the conversion of a deep sea into a shoal, some small change in organic life may have taken place, and consequently the distinctness in character of the fossils of the two formations may be derived from two causes, first and principally the difference of geographical conditions, and, secondly, that law of the coming in and going out of species which was alluded to in the last chapter. (p. 275.)

The area over which both divisions of the crag can be traced is too small to enable us to arrive at satisfactory conclusions on a question of such magnitude; but the section given above (p. 302.) shows distinctly that, near Sutton, the lower crag had suffered much denudation before the deposition of the red crag. At D (Fig. 128.) there is not only a distinct cliff, eight or ten feet high, of coralline crag, running in a direction N. E. and S. W., against which the red crag abuts with its horizontal layers, but this cliff occasionally overhangs. The rock composing it is drilled everywhere by Pholades belonging to the period of the red crag. The cliff may have been caused by submarine denudation, in a shallow sea; and had the red crag been equally solid, it would probably have presented many similar perpendicular cliffs; for beds, ten or twelve feet thick, of loam or sand, in this formation, are often seen to be unconformable to older beds, which have been in part cut away. Similar excavations are now made, even on a larger scale, by the sea, in the great sandbanks off Yarmouth, in part of which Captain Hewett, R. N., found, in 1836, a broad channel, sixty-five feet deep, where there had been only a depth of four feet in 1822. This remarkable change was ascertained during two hydrographical surveys, in the years above mentioned, and shows how denudation, amounting to sixty feet in vertical depth, can take place under water in the course of fourteen years. The new channel thus formed, serves now (1838) for the entrance of ships into Yarmouth Roads.

Eocene formations in England—London Clay. — In the section already given of the tertiary strata of Suffolk (p. 302.), it will be seen that the crag rests on a formation called the London clay, which there consists of alternating beds of blue and brown clay, with many nodules of calcareous stone, used for Roman cement. This formation is well seen in the neighbouring cliffs of Harwich, where the nodules contain many marine shells, and sometimes the bones of turtles. The relative position

Fig. 136.

of the chalk, London clay, and crag, between the coast of Essex and the interior, may be understood by reference to the annexed diagram. The London clay has been so named, because it occurs in the neighbourhood of the metropolis, in a trough or basin of the chalk. (Sèe section, p. 315.) We know, by numerous borings made for water, that the chalk exists everywhere below, after we have penetrated through clay and sand to the depth of from 200 to 600 feet; and, if we proceed to the south of London, we find the chalk rising up to the surface and forming the Surrey hills; while if we proceed northwards, into Hertfordshire, or, westward, by the Thames, into Oxfordshire, we again meet with the same chalk.

The overlying Eocene deposit consists of two portions; the upper of blue clay, with occasional cement stones, as before mentioned; the lower of various coloured sands and clays; the fossils throughout all the beds being very different from those of the crag. Scarcely any one of the shells can be identified with species now living; and the whole assemblage is such as to resemble the testaceous fauna of the tropics. This opinion is favoured by the occurrence of many species of Mitra and Voluta, a large Cypræa, a very large Rostellaria, and shells of the genera Terebellum, Cancellaria, Crassatella, and others, with four or more species of Nautilus. (See Figures, p. 310.) There are fish, also, which indicate a warm climate; among which may be mentioned a sword-fish, (*Tetrapterus priscus*, Agassiz,) about eight feet long, and a saw-fish, (*Pristis bisulcatus*, Ag.) about ten feet in length; genera foreign to the British seas.

These last have been found in the island of Sheppey, which is composed of London clay, where

FOSSIL SHELLS OF THE LONDON CLAY.

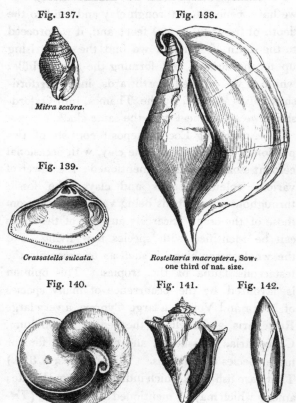

Fig. 137.　　　　　　　　　　　Fig. 138.

Mitra scabra.

Fig. 139.

Crassatella sulcata.

Rostellaria macroptera, Sow. one third of nat. size.

Fig. 140.　　　　　Fig. 141.　　　Fig. 142.

Nautilus centralis.　　*Voluta athleta.*　　*Terebellum fusiforme.*

also, as I learn from M. Agassiz, the remains of no less than fifty other species of fish have been discovered. It does not appear that the fossil plants and fruits, so numerous in this same island, or the fossil plants of the corresponding Eocene formation of Paris, have by any means so tropical an aspect as the shells, but rather indicate such a flora as might be found on the borders of the Mediterranean.

Besides the marine formation called London clay, there are freshwater strata of the Eocene period in the Isle of Wight, and opposite coast of Hampshire. They contain shells, such as Limnea and Planorbis; Gyrogonites, or the fossil seeds of Chara (see p. 66.); and the bones of several quadrupeds of extinct genera, such as Palæotherium, Anoplotherium, and Chæropotamus, which were lately found by the Rev. W. D. Fox, near Binstead.

It has already been remarked that fossil mammalia of extinct species have been met with in the Newer Pliocene deposits of England and other countries. Different species characterize the Miocene, and others are proper to the Eocene formations; and among them nearly every order and family of the herbivorous and carnivorous tribes are represented: but those which inhabit trees are most rare; and it was not until very lately, namely in 1837, that any remains of quadrumana, or of the ape and monkey tribe, were discovered. These were obtained about the same time in France and India; in France, by M. Lartet,

near Auch, in the department of Gers, about forty
miles west of Toulouse, where the bones of an ape,
or gibbon, accompanied those of the rhinoceros,
dinotherium, mastodon, and others; in India, by
Captain Cautley and Dr. Falconer, who found the
remains of a monkey, with the bones of many extinct quadrupeds, in the Sewalik hills, a lower
range of the Himalaya mountains, near Saharun-
pore.

The frequent occurrence in the tertiary strata
of fossils referable to the highest class of vertebrata is a fact the more worthy of notice, as we
shall find in the sequel how great is their rarity
in the secondary formations.

CHAPTER XV.

CRETACEOUS GROUP.

White chalk — Its marine origin shown by fossil shells — Extinct genera of cephalopoda — Sponges and corals in the chalk — No terrestrial or fluviatile shells, no land plants — Supposed origin of white chalk from decomposed corals — Single pebbles, whence derived — Cretaceous coral-reef in Denmark — Maestricht beds and fossils — Origin of flint in chalk — Wide area covered by chalk — Green-sand formation and fossils — Origin of — External configuration of chalk — Outstanding columns or needles — Period of emergence from the sea — Difference of the chalk of the north and south of Europe — Hippurites — Nummulites — Altered lithological character of cretaceous formation in Spain and Greece — Terminology.

THE next group which succeeds to the tertiary strata in the descending order has been called Cretaceous or chalky, because it consists in part of that remarkable white earthy limestone called chalk (*creta*). With this limestone however are usually associated other deposits of sand, marl, and clay, called the Green-sand formation, because some of its sands are remarkable for their bright green colour.

The following are the subdivisions into which the Cretaceous Strata have been divided in the south of England: —

P

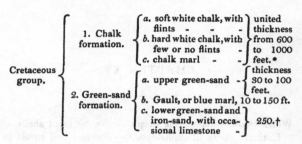

The accompanying section (Fig. 143.) will show the manner in which the tertiary strata of the London and Paris basins, as they are called, rest upon the chalk, and how the white chalk in its turn reposes throughout this region upon the green-sand formation.

I shall now speak first of the chalk, its fossils, and probable origin; and then say something of the green-sand; after-which I shall point out the probable relations of the chalk and green-sand to each other.

White Chalk. — The white chalk used in writing consists almost purely of carbonate of lime. Although usually soft, this substance passes in some districts by a gradual change into a solid stone used for building. The stratification is often obscure, except where rendered distinct by alternating layers of flint. These layers are from two to four feet distant from each other, and from three

---

* Conybeare, Outlines, &c., p. 85.

† Fitton, Geol. Trans., Second Series, vol. iv. p. 319.

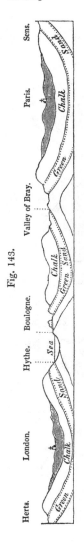

Fig. 143.

Section from Hertfordshire, in England, to Sens, in France.

to six inches in thickness, occasionally in continuous beds, but more frequently in nodules.

The annexed figures represent some few of the fossil shells which are abundant in the white chalk, and these alone are sufficient to prove its marine origin. Some of them, such as the Terebratulæ, (see Figs. 148. 150, 151, 152.) are known to live at the bottom of the sea, where the water is tranquil, and of some depth. The Crania and Catillus (Figs. 145. & 144.) may be pointed out as forms which, so far as our present information extends, became extinct at the close of the cretaceous period, and are therefore never met with in any tertiary stratum, or in a living state. Among other forms, equally conspicuous among the fossil mollusca of the cretaceous group, and foreign to the tertiary and recent periods, may be mentioned the Belemnite, Ammonite, Baculite, and Turrilite of the family Cephalopoda, to which the living cuttle-fish and nautilus belong.

FOSSILS OF THE WHITE CHALK.

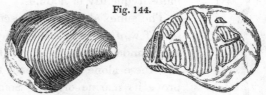

Catillus Cuvieri. (Syn. *Inoceramus Cuvieri*, Sow.)

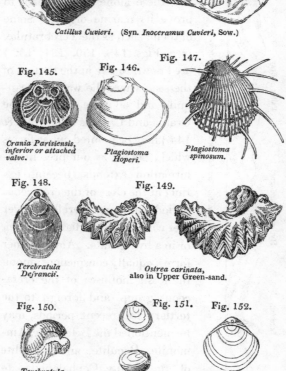

Fig. 145.

*Crania Parisiensis,
inferior or attached
valve.*

Fig. 146.

*Plagiostoma
Hoperi.*

Fig. 147.

*Plagiostoma
spinosum.*

Fig. 148.

*Terebratula
Defrancii.*

Fig. 149.

*Ostrea carinata,
also in Upper Green-sand.*

Fig. 150.

*Terebratula
octoplicata,*
(*Var. of T. plicatilis.*)

Fig. 151.

Terebratula pumilus.
(*Magas pumilus*, Sow.)

Fig. 152.

*Terebratula
carnea.*

Fig. 153.

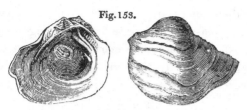

Ostrea vesicularis (Gryphæa globosa, Sow.)
also in Upper Green-sand.

FOSSIL CEPHALOPODA OF EXTINCT GENERA.

Cretaceous Period.

a   Fig. 154.   *b*

a. *Turrilites costatus,* Gault.
b. *Same, showing the indented border of the partition of the chambers.*

a   Fig. 155.   *b*

a. *Belemnites mucronatus,* } White Chalk and Upper Green-sand.
b. *Same, internal structure,* }

Fig. 156.

Fig. 157.

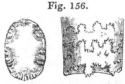

Portion of Baculites Faujasii,
White Chalk and Upper Green-sand.

Portion of Baculites anceps,
White Chalk and Upper Green-sand.

One of these, the Belemnite, like the bone of
the common cuttle-fish, was an internal shell. Be-

sides these there are other fossils in the chalk,
such as sea-urchins, corals, and sponges (see
Figures), which are alike marine.   They are dispersed indifferently through the soft chalk and
the hard flint.

Fig. 158.

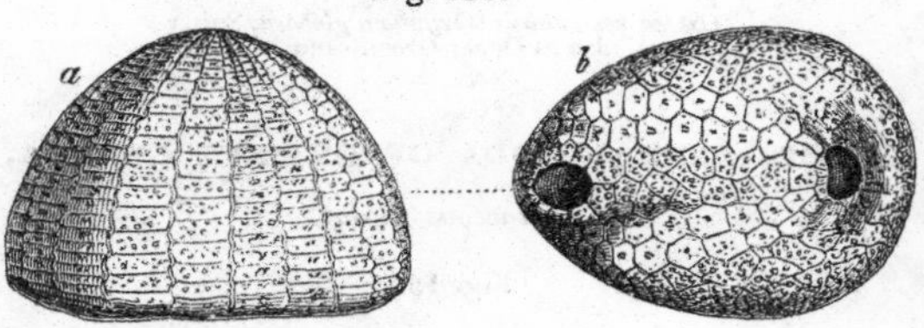

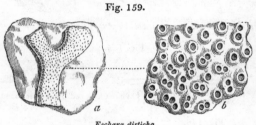

Ananchytes ovatus.

a. side view.
b. bottom of the shell on which both the oral and anal apertures
    are placed ; the anal being to the right, and more oval.

Fig. 159.

Eschara disticha.
a. natural size.   *b.* portion magnified.

To some of these inclosed zoophytes many flints
owe their irregular forms, as in the flint represented in Fig. 161., where the hollows on the exterior are caused by the branches of a sponge,
which is seen on breaking open the flint. (See
Fig. 160.)

Fig. 160.   Fig. 161.

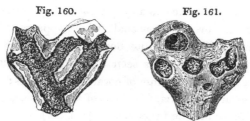

*A branching sponge in a flint from the chalk.**

With these fossils the remains of fish and crustacea are not uncommon; but we meet with no bones of land animals, nor any terrestrial or fluviatile shells, nor any plants, except pieces of drift wood and sea-weed, nor any sand or pebbles; all the appearances concur in leading us to believe that this deposit was formed in a deep sea, far from land, and at a time when the European fauna was perfectly distinct from that of the tertiary period, from which its numerous species of plants and animals entirely differ.

Origin of the White Chalk. — Having then come to the conclusion, that the chalk was formed in an open sea of some depth; we may next inquire, in what manner so large a quantity of this peculiar white substance could have accumulated over an area many hundred miles in diameter, and some of the extreme points of which are distant, as we shall see in the sequel, more than 1000 geographical miles from each other.

* From the collection of Mr. Bowerbank.

It was remarked in an early part of this volume, that some even of that chalk which appears to an ordinary observer quite destitute of organic remains, is nevertheless seen under the microscope to be full of fragments of corals and sponges; the valves of Cytherina, the shells of foraminifera, and still more minute infusoria. (See p. 55.)

Now it had been often suspected before these discoveries, that white chalk might be of animal origin, even where every trace of organic structure has vanished. This bold idea was partly founded on the fact, that the chalk consisted of pure carbonate of lime, such as would result from the decomposition of testacea, echini, and corals, and in the passage observable between these fossils when half decomposed into chalk. But this conjecture seemed to many naturalists quite vague and visionary, until its probability was strengthened by new evidence brought to light by modern geologists.

We learn from Lieutenant Nelson, that, in the Bermuda islands, there are several basins or lagoons almost surrounded and inclosed by reefs of coral. At the bottom of these lagoons a soft white calcareous mud is formed by the decomposition of Eschara, Flustra, Cellepora, and other soft corallines. This mud, when dried, is undistinguishable from common white earthy chalk; and some portions of it, presented to the Museum of the Geological Society of London, might, after full examination, be mistaken for ancient chalk, but for the labels attached to them. About the same time Mr. C. Darwin observed similar facts in the coral islands of the Pacific; and came also to the opinion, that much of the soft white mud found at the bottom of the sea near coral reefs has passed through the bodies of worms, by which the stony masses of coral are everywhere bored; and other portions through the intestines of fish; for certain gregarious fish of the genus *Sparus* are visible through the clear water, browsing quietly, in great numbers, on living corals, like grazing herds of graminivorous quadrupeds. On opening their bodies, Mr. Darwin found their intestines filled with impure chalk. This circumstance is the more in point, when we recollect how the fos-silist was formerly puzzled by meeting with cer-

Fig. 162.   Fig. 163.

Coprolites of fish called Tulo-
cido-copri, from the chalk.

tain bodies, called cones of the larch, in chalk, which were afterwards recognized by Dr. Buckland to be the excre-ment of fish.* These spiral coprolites (see Figures) like the scales and bones of fossil fish in the chalk, are composed chiefly of phosphate of lime.

* Geol. Trans., Second Series, vol. iii. p. 232. plate 31. figs. 3. and 11.

Single pebbles in chalk. — The general absence of
sand and pebbles in the white chalk has been already
mentioned; but the occurrence here and there of
a few isolated pebbles of quartz and green-schist,
some of them two or three inches in diameter, in
the south-east of England, has justly excited much
wonder. If these had been carried to the spots
where we now find them by waves or currents
from the lands once bordering the cretaceous sea,
how happened it that no sand or mud were transported thither at the same time? We cannot
conceive such rounded stones to have been drifted
like erratic blocks by ice *, for that would imply
a cold climate in the cretaceous period; a sup-
position inconsistent with the luxuriant growth of
large chambered univalves, numerous corals, and
many fish, and other fossils of tropical forms.

Now in Keeling Island, one of those detached
masses of coral which rise up in the wide Pacific,
Captain Ross found a single fragment of greenstone, where every other particle of matter was
calcareous; and Mr. Darwin concludes that it
must have come there entangled in the roots of a
large tree. He reminds us that Chamisso, a distinguished naturalist who accompanied Kotzebue,
affirms, that the inhabitants of the Radack archipelago, a group of lagoon islands, in the midst of

* See p. 136.

the Pacific, obtained stones for sharpening their instruments by searching the roots of trees which are cast up on the beach.*

It may perhaps be objected, that a similar mode of transport cannot have happened in the cretaceous sea, because fossil wood is very rare in the chalk. Nevertheless wood is sometimes met with, and in the same parts of the chalk where the pebbles are found, both in soft stone and in a silicified state in flints. In these cases it has often every appearance of having been floated from a distance, being usually perforated by boring-shells, such as the Teredo and Fistulana. ‡

The only other mode of transport which suggests itself is sea-weed. Dr. Beck informs me, that in the Lym-Fiord, in Jutland, the *Fucus vesiculosus*, sometimes grows to the height of ten feet, and the branches rising from a single root, form a cluster several feet in diameter. When the bladders are distended, the plant becomes so buoyant as to float up loose stones several inches in diameter, and these are often thrown by the waves high up on the beach. The *Fucus giganteus*, of Solander, so common in Terra del Fuego, is said by Captain Cook to obtain the length of 360 feet, although the stem is not much thicker than a

* Darwin, p. 549.   Kotzebue's First Voyage, vol. iii. p. 155.

† Mantell, Geol. of S. E. of England, p. 96.

man's thumb. It is often met with floating at
sea, with shells attached, several hundred miles
from the spots where it grew. Some of these
plants, says Mr. Darwin, were found adhering to
large loose stones in the inland channels of Terra
del Fuego, during the voyage of the Beagle in
1834; and that so firmly, that the stones were
drawn up from the bottom into the boat, although
so heavy that they could scarcely be lifted in by
one person. * Some fossil sea-weeds have been
found in the cretaceous formation, but none, as
yet, of large size.

Cretaceous coral reef in Denmark.— Having said
so much on the probable derivation of chalk from
the decay of corals and shells, I may add, that in
the island of Seeland, in Denmark, there is a yellow limestone intimately connected with the chalk,
and containing a vast number of the same fossils,
which consists of an aggregate of corals, retaining
their forms as distinctly as the dead zoophytes
which enter into the structure of reefs now growing in the sea. The thickness of this rock is unknown, but it has been quarried at Faxoe to the
depth of forty feet. At Stevensklint, in Seeland,
it is seen to rest on white chalk with flints, from
which it differs greatly in appearance, and where
it is covered again by another limestone, which

* Darwin, p. 303. (For full reference see p. 137.)

although of later date, agrees more nearly with the white chalk, both in fossils and mineral character. Out of 104 species of sponges, corals, and other zoophytes, collected from the limestone of Faxoe, and from the ordinary white chalk of Denmark, which agrees with that of England, no less than forty-two are common to both formations; and many of the same species of bivalve shells and echinodermata have been found in both. The Faxoe formation, however, is not only remarkable for the number and good preservation of its fossil corals, but also from the generic resemblance of many of its univalve shells to forms usually supposed to appertain chiefly or exclusively to the tertiary period. Thus among the patelliform univalves, we find Patella and Emarginula, and among the spiral, the following genera, Cypræa, Oliva, Mitra, Cerithium, Fusus, Trochus, Triton, Nassa, and Bulla.

The *species* however do not agree with those of the tertiary strata, and are associated with cephalopoda of those extinct families before mentioned as characteristic of the cretaceous, and foreign to the tertiary epoch; as, for example, the ammonite, belemnite, and baculite. Two species, the *Belemnites mucronatus* (Fig. 155.), and the *Baculites Faujasii* (Fig. 156.), being common to the Faxoe beds and the white chalk.

From these facts, we may conclude that the

Faxoe limestone was formed in the cretaceous sea, in a spot favourable for the multiplication of stony corals and univalve shells; and as some small portions of the rock consist of white earthy chalk, this latter substance must have been produced simultaneously, and some of it may have been washed away, in the form of mud, from the coral reef of Faxoe, and dispersed over the deeper parts of the same ocean, just as the white mud, swept out of the lagoons of the Bermudas or coral islets of the Pacific, must form deposits of white chalk, covering much wider spaces than those occupied by the reefs.

The same remarks apply to a rock, which reposes on the Upper Chalk with flints, at St. Peter's Mount, Maestricht, and at Ciply, near Mons. It is a soft yellowish stone, not very unlike chalk, and " includes siliceous masses, which are much more rare than those of the chalk, of greater bulk, and not composed of black flint, but of chert and calcedony." * Like the Faxoe stone, it is characterized by a peculiar assemblage of organic remains which are specifically distinct from those of the tertiary period, but many of them common to the white chalk.

As these Maestricht beds have been thought to be intermediate in character between the second-

* Fitton, Geol Proceedings, 1830.

ary and tertiary formations, it may be proper to
mention, as opposed to this opinion, that the Ammonite (Fig. 164.), Baculite, Hamite, and Hip-

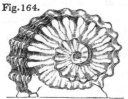

Fig.164.

Ammonites Rhotomagensis,
Maestricht; found by Count
Munster.

purite, have been found in
the Maestricht limestone,
genera which have not yet
been detected in strata
newer than the chalk. In
the same formation, also,
large turtles have been found,
and a gigantic reptile, the
Mosasaurus, or fossil Monitor, some of the vertebræ of which appear also in the English chalk. *
The osteological characters of this oviparous quadruped prove it to have been intermediate between the living Monitors and Iguanas; and, from
the size of the head, vertebræ, and other bones,
it is supposed to have been twenty-four feet in
length.

The existence of such turtles and saurians
seems to imply some neighbouring land, on the
sandy shores of which these creatures may have
laid their eggs. But a few small islets in mid
ocean, like Ascension, so much frequented by
turtles, may perhaps have afforded the required
retreat to these cretaceous reptiles.

Origin of the flint in chalk. — It is difficult to

* See Mantell's Geol. of S. E. of England.

give a satisfactory explanation of the origin of the
flint in chalk, whether it occurs in nodules or
continuous layers.   It seems that there was originally siliceous as well as calcareous earth in the
muddy bottom of the cretaceous sea, at least when
the upper chalk was deposited.   Whether both
these earths could have been alike supplied by
the decay of organic bodies may be matter of
speculation ; but what was said of the origin of
Tripoli (see p. 51.) shows how microscopic infusoria can give rise to dense masses of pure flint.
The skeletons of many living sponges consist of
needles or spicula of flint, and these are found
very abundantly in the flints of the chalk.   There
are also other living zoophytes, which have the
power of secreting siliceous matters from the
waters of the sea, just as mollusca secrete calcareous particles.

From whatever source the mud derived its
silex, we may attribute the parallel disposition of
the flinty layers to successive deposition.   The
distances between the layers, says Dr. Buckland,
must have been regulated by the intervals of pre-
cipitation, each new mass forming at the bottom
of the ocean a bed of pulpy fluid, which did not
penetrate the preceding bed on which it rested,
because the consolidation of this last was so far
advanced as to prevent such intermixture.*   Never-

* Geol. Trans., First Series, vol. iv. p. 420.

theless the separation of the flint into layers, so distinct from the chalk, is a singular phenomenon, and not yet accounted for. Perhaps, as the specific gravity of the siliceous exceeds that of the calcareous particles, the heavier flint may have sunk to the bottom of each stratum of soft mud?

Geographical extent of White Chalk. — The area over which the white chalk preserves a nearly homogeneous aspect is so great that geologists have often despaired of finding any analogous deposits of recent date; for chalk is met with in a north-west and south-east direction, from the north of Ireland to the Crimea, a distance of about 1140 geographical miles, and in an opposite direction it extends from the south of Sweden to the south of Bordeaux, a distance of about 840 geographical miles. But we must not conclude that it was ever spread out uniformly over the whole of this vast space, but merely that there were patches of it, of various sizes, throughout this area. Now, if we turn to those regions of the Pacific over which coral reefs are scattered, we find some archipelagoes of lagoon islands, such as that of the Dangerous archipelago for instance, and that of Radack, with some adjoining groups, which are from 1100 to 1200 miles in length, and 300 or 400 miles broad; and the space to which Flinders proposed to give the name of the Corallian sea is still larger; for it is bounded on the east by the Australian barrier, on the west by New Caledonia, and on the north by the reefs of Louisiade. Although the islands in these spaces may be thinly sown, the mud of the decomposing zoophytes may be scattered far and wide by oceanic currents.

Green-sand formation. — The lower division of the Cretaceous group in England is divisible, as we have already seen, into Upper Green-sand, Gault, and Lower Green-sand. The green grains have been found, by analysis, to consist chiefly of silicate of iron, and they agree in composition with chlorite. The inferior white marly chalk becomes more and more charged with these grains until it passes into the upper green-sand, a formation of sand and sandy marl, frequently mixed with chert, and this again passes downwards into the clay and marl, provincially called Gault. Both of these subdivisions, although often diminishing in volume to a thickness of two or three yards, form distinct and continuous bands of sand and clay between the chalk and lower green-sand throughout considerable tracts in England, France, and Belgium; and each preserve throughout this space certain mineral peculiarities and characteristic fossils.

The lower green-sand below the gault is formed partly of green and partly of ferruginous sand and sandstone, with some limestone. These rocks

succeed each other in the following descending
order in Kent : —

No. 1. Sand, white, yellowish, or ferruginous, with concretions
         of limestone and chert   -   - 70 feet.
     2. Sand with green matter   -   - 70 to 100 feet.
     3. Calcareous stone, called Kentish rag   - 60 to 80 feet.*

The fossils of the green-sand are marine, and
some of them, like the *Pecten quinquecostatus*
(Fig. 166.), range through all the members of the

FOSSILS OF THE GREEN-SAND FORMATION.

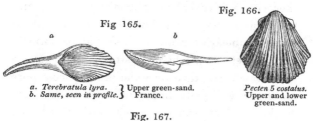

Fig 165.

Fig. 166.

a. *Terebratula lyra.* } Upper green-sand.
b. *Same, seen in profile.* } France.

Pecten 5 costatus.
Upper and lower
green-sand.

Fig. 167.

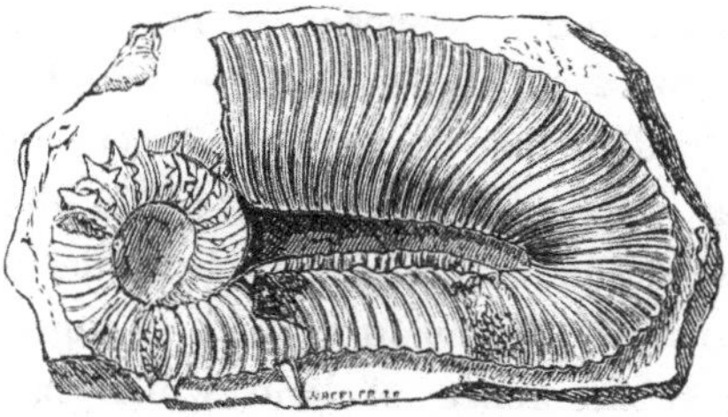

Hamites spiniger (Fitton), near Folkstone.†

series.   Several forms of cephalopoda, such as the
Hamite (Fig. 167.), Scaphite, and others distin-

* Fitton, Geol. Trans., Second Series, vol. iv. p. 319.
† Ibid. pl. 12.

guish the Green-sand formation in England from the White Chalk.

Origin of the Green-sand formation. — Unlike the white chalk, this deposit consists of a succession of ordinary beds of sand, clay, marl, and impure limestone, the materials of which might result from the wearing down of pre-existing rocks. The nature of these derivative rocks we learn, from finding in the green-sand pebbles of quartz, quartzose sandstone, jasper, and flinty slate, together with grains of chlorite and mica.* But we naturally inquire, how it could happen that, throughout a large submarine area, there should be formed, first, a set of mechanical strata, such as the green-sand, and then over the same space a pure zoophytic and shelly limestone, such as the white chalk. Certain causes, which during the first period gave rise to deposits of mud, sand, and pebbles, must subsequently have ceased to act; for it is evident that no similar sediment disturbed the clear waters of the sea in which the white chalk accumulated. The only hypothesis which seems capable of explaining such changes is the gradual submergence of land which had been previously exposed to aqueous denudation. This operation may have gone on with such slowness as to allow time for considerable fluctuations in the

* Fitton, Geol. Trans., Second Series, vol. iv. p. 116.

state of the organic world, so that different sets of strata, beginning with the lower green-sand, and ending with the upper white chalk, may each contain some peculiar remains of animals which lived successively in the sea; while some species may have continued to exist throughout the whole period, and are therefore common to all these formations.

It will be seen in the next chapter, when we treat of the strata called the Wealden, that such a general subsidence of land as is here supposed to explain the manner in which the chalk succeeds the green-sand, may be inferred from other independent proofs to have taken place throughout large areas.

It cannot however be assumed, that all the green-sand in Europe had ceased to be deposited before any chalk began to accumulate. Such indeed was the order of events in parts of England, France, Belgium, and Denmark; but if we compare different countries, and some of these not far distant from each other, we find reason to believe that sand and clay continued to be thrown down in one place, while pure chalk was forming in another. In Westphalia, for example, strata containing the same fossils as the white chalk of England, consist of sand and marl with green grains like the upper green-sand. Similar facts have been observed in Hungary in the Carpathian mountain chain. Such variations would occur if the supposed sinking down of land did not take place simultaneously everywhere; and for this reason the minor subdivisions of the cretaceous group, however persistent and uniform in their mineral characters in some regions, vary rapidly, and change entirely in other directions.

External configuration of Chalk. — The smooth rounded outline of the hills composed of white chalk is well known to all who have travelled in the south-east of England. The chalk downs, being free from trees or hedge-rows, afford us an opportunity of observing the manner in which the upper valleys unite with larger ones, and how these become wider and deeper as they descend. For the most part they are dry, yet occasionally they afford a perfect system of drainage, when a sudden flood is caused by heavy rains or the melting of snow. We may conceive their excavation to have been caused by the action of the waves and currents while the chalk was gradually emerging from the sea. To the same action we may ascribe the escarpments as they are called, or those long lines of precipitous cliffs in which the chalk often terminates abruptly, and which, though now inland, have been undermined by the waves when the chalk was upheaved from the sea.

Many examples occur in England; but there are no precipices of chalk more striking than those which bound the lower part of the great

valley or gorge through which the Seine flows in
Normandy. At various heights on the steep sides
of these hills are outstanding pillars and pinnacles
of a very hard and compact chalk, as at Tournedos and Elbeuf, near Rouen, which evidently owe
their shape to the power of the waves. (See Figures.)

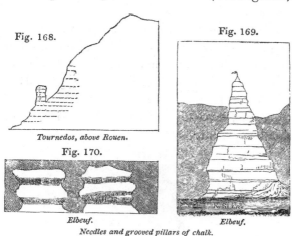

Fig. 168.

Tournedos, above Rouen.

Fig. 169.

Fig. 170.

Elbeuf.

Elbeuf.

Needles and grooved pillars of chalk.

Some small columns near Elbeuf exhibit parallel
and horizontal grooves scooped out of the columns
at different heights. (See Fig. 170.) These greatly
resemble certain limestone pillars, described by
Captain Bayfield, in the Mingan islands in the
gulf of St. Lawrence. There is evidence there
of the coast having been upheaved at successive
periods, so that parallel ranges of sea beaches,
with recent shells, have been laid dry, terrace above

terrace.   At heights corresponding to the beaches
the isolated masses of calcareous rock retain the
marks worn by the waves.   These marks probably
indicate pauses in the upheaving process, during
which the sea had a considerable time to wear
away the stone as well as to throw up a beach at
the same level.*

The needles of the Isle of Wight, and the Old
Harry Rocks of the coast of Dorsetshire, are well
known to those who have examined the chalk
cliffs of the South of England.   Besides the inland
columns in Normandy, above described, there are
others more recently formed on the sea coast of
that same country.

Fig. 171.

*Needle and Arch of Etretat, in the chalk cliffs of Normandy.
Height of Arch 100 feet.   (Passy.)* †

If we inquire at what period the emergence
and denudation of the cretaceous rocks took place,

* Captain Bayfield, Geol. Trans., Second Series, vol. v.
p. 94.   Also Princ. of Geol., Index, " Niapisca island."
† Seine-Inferieure, p. 142. and plate 6. fig. 1.

we shall find that it occurred in great part after the deposition of various marine tertiary formations, so that both the cretaceous and tertiary beds were upraised together.   The greatest elevation which the chalk reaches in England, is the summit of Inkpen Beacon in Berkshire, which is 1011 feet above the sea; but marine deposits of the same age attain an elevation of 8000 feet in the Alps and Pyrenees.   These may have partly emerged during the cretaceous period, just as the coral reefs in some regions of the Pacific are growing in one spot, while other portions of the same have been uplifted by subterranean forces, and converted into land.

Difference between the chalk of the north and south of Europe. — By the aid of the three tests of relative age, namely, superposition, mineral character, and fossils, the geologist has been enabled to refer to the same cretaceous period certain rocks in the north and south of Europe, which differ greatly both in their fossil contents, and in their mineral composition and structure.

If we attempt to trace the cretaceous deposits from England and France to the countries bordering the Mediterranean, we perceive, in the first place, that the chalk and green-sand in the neighbourhood of London and Paris, form one great continuous mass, the strait of Dover being a trifling interruption, a mere valley with chalk cliffs on both sides. We then observe that the main body of the chalk which surrounds Paris stretches from Tours to near Poitiers, (see the annexed map (Fig. 172.), in which the shaded part represents chalk.)

Fig. 172.

Between Poitiers and La Rochelle, the space marked A on the map separates two regions of chalk. This space is occupied by the oolite and certain other formations older than the chalk, and has been supposed by M. E. de Beaumont to have formed an island in the cretaceous sea. South of this space we again meet with a formation which we at once recognize by its mineral character to be chalk, although there are some places where the rock becomes oolitic. The fossils also are upon the whole very similar, although some new forms now begin to appear in abundance which are rare or wholly unknown further to the north. Among these may be mentioned many Hippurites, Sphærulites, and other members of that great family of mollusca called *Rudistes* by

Lamarck, to which nothing analogous has been discovered in the living creation. Although very uncommon in England, one species of this family has been discovered in our chalk.

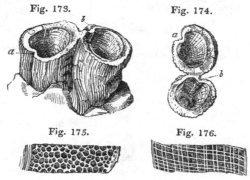

Fig. 173.   Fig. 174.

Fig. 175.   Fig. 176.

Hippurites Mortoni, Mantell.   Maidstone, Kent.
Diameter one-seventh of nat. size.

Fig. 173. Two individuals deprived of their opercula, adhering
together.
    174. Same seen from above.
    175. Transverse section of part of the wall of the shell,
magnified to show the structure.
    176. Vertical section of the same.

On the side where the shell is thinnest, there is one external furrow and corresponding internal ridge, *a. b.* Figs. 173, 174. ; but they are usually less prominent than in these figures. This species has been referred to Hippurites, but does not, I believe, fully agree in character with that genus. I have never seen the opercular piece, or *valve,* as it is called by those conchologists who regard the *Rudistes* as bivalve mollusca.

But this family, which is so feebly represented in England and the north of France, becomes quite characteristic of rocks of the cretaceous era in the south of France, Spain, Greece, and other countries bordering the Mediterranean.

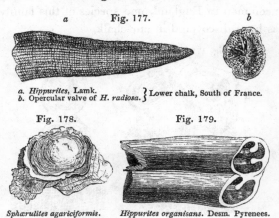

a   Fig. 177.   *b*

a. *Hippurites*, Lamk.
b. Opercular valve of *H. radiosa.* } Lower chalk, South of France.

Fig. 178.   Fig. 179.

Sphærulites agariciformis.   *Hippurites organisans.* Desm. Pyrenees.

Between the region of chalk last mentioned in which Perigeux is situated, and the Pyrenees, the space B intervenes (see Map).

Here the tertiary strata cover, and for the most part conceal, the cretaceous rocks, except in some spots where they have been laid open to view by the denudation of the newer formations. In these places they are seen still preserving the form of a white chalky rock, which is filled in part with grains of green sand. Even as far south as Tercis, on the Adour, near Dax, it retains this character where I have examined it, and where M. Grateloup has found in it *Ananchytes ovata* (Fig. 158.), and other fossils of the English chalk, together

with Hippurites. When we arrive at Bayonne
and the Pyrenees, the cretaceous formation, although still exhibiting some of the same mineralogical peculiarities, is nevertheless greatly changed.
Its calcareous division consists for the most part of
compact crystalline marble, often full of nummulites (see Fig. 180.), and those portions which
may be imagined to represent the green-sand, are
composed of shales, grits, and micaceous sandstone, containing impressions of marine plants,
together with lignite and coal. There are also
beds of red sandstone and conglomerate belonging
to the same group. These rocks ascend gradually
into the highest parts of the Pyrenees, and cross
over into Spain, where the cretaceous system assumes a character still more unlike that of northern
Europe.

Here, as on the north side of the Pyrenees, the
most conspicuous fossils are hippurites, sphæru-

Fig. 180.

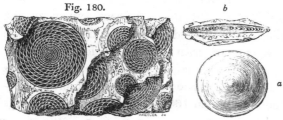

Nummulite limestone; Peyrehorade, Pyrenees.

a. External surface of one of the nummulites, of which longitudinal
   sections are seen in the limestone.
b. Transverse section of same.

Q 3

lites, and nummulites.   The last-mentioned fossil, so called from its resemblance to a piece of money, is a genus of mollusca very abundant in the tertiary strata of Northern Europe; but only met with in chalk in the South of Europe.

So many species and genera of shells now wanting in our northern seas, are frequent in the Mediterranean, that we need not be surprised, when following from north to south the deposits of the old cretaceous sea, at finding similar modifications in organic forms.

The cretaceous rocks in the Alps, Italy, Greece, and Asia Minor, are distinct in like manner from the type of that formation in the North of Europe; yet their age in most of these countries can be clearly ascertained, partly by following them continuously from the north in the manner above described; and partly by their position below the tertiary, and above the oolitic strata.

We learn from the researches of M. M. Boblaye and Virlet, that the cretaceous system in the Morea, is composed of compact and lithographic limestones of great thickness; also of granular limestones, with jasper; and in some districts, as in Messenia, a puddingstone with a siliceous cement more than 1600 feet in thickness, belongs to the same group.*

* Bull. de la Soc. Géol. de France, tom. iii. p. 149.

It is evident, observe these geologists, from the great range of the hippurite and nummulite limestone, that the South of Europe was occupied at the cretaceous period by an immense sea, which extended from the Atlantic Ocean into Asia, and comprehended the southernmost part of France, together with Spain, Sicily, part of Italy, and the Austrian Alps, Dalmatia, Albania, a portion of Syria, the isles of the Ægean, coasts of Thrace, and the Troad.

In proportion, therefore, as we enlarge the sphere of our researches, we may find in the strata of one era, the mineralogical counterparts of the rocks, which, in a single country like England, may characterise successive periods. Thus, the grits, sandstone, and shale with coal, of the Pyrenees have actually been mistaken by skilful miners for the ancient carboniferous group of England and France. In like manner the cretaceous red marl and salt of northern Spain have been regarded as the same as our new red and saliferous sandstone; and the lithographic limestone of the Morea might be confounded with the oolite of Solenhofen in Germany.

The beginner, perhaps, on hearing these facts, may object to the term cretaceous, as applied to the rocks of the southern region in which there is no chalk. But the term green-sand would have been equally inappropriate as a general name for this group; and that of hippurite and nummulite limestone, however well suited to the Mediterranean region, would be inapplicable to the chalk of the north. Scarcely any designation would remain unexceptionable as we enlarge the bounds of our knowledge, and we must therefore be content to retain many ancient names, as simply expressing the mineral, or palæontological characters of rocks *in the country where they were first studied*.

CHAPTER XVI.

BENEATH the cretaceous rocks in the S. E. of
England, a freshwater formation is found called
the Wealden, which, although it occupies a small
area in Europe, as compared to the chalk, is nevertheless of great interest, as being intercalated between two marine formations. It is composed of
three minor groups, of which the aggregate thickness in some places cannot be less than 800 feet.*
These subdivisions are,

	Thickness.
1st. Weald clay, sometimes including thin beds of sand and shelly limestone	140 to 280 ft.
2d. Hastings sand, in which occurs some clays and calcareous grits ;—between	400 and 500 ft.
3d. Purbeck beds, consisting of various kinds of limestones and marls	about 250 ft.

* Dr. Fitton, Geol. Trans. vol. iv. p. 320. Second Series.

Q 5

To all these subdivisions, the common name of the Wealden has been given, because they may be best studied in part of Kent, Surrey, and Sussex, called the Weald.

We have seen that the fossils of the chalk and green-sands which repose upon the Wealden are all marine, and the species numerous; and the same remark applies to the Portland stone and other members of the Oolitic series which lie immediately beneath (see Fig. 181.). But in the

Fig. 181.

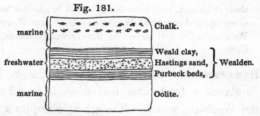

Position of the Wealden between two marine formations.

Wealden itself, although the fossils are abundant as to quantity, the number of different species is comparatively small, and by far the greater part of them show that they were deposited in a freshwater lake, or estuary communicating with the sea.*

Fossils of the Wealden. — The shells of this formation are almost exclusively of fluviatile or lacustrine genera, such as Melanopsis, Paludina, Neritina, Cyclas, Unio, and others. The individuals are sometimes in such profusion, that the

* Fitton, Geol. Trans. vol. iv. p. 104. Second Series.

surface of each thin layer of marl or clay is covered
with the valves of Cyclas, and whole beds of limestone are almost entirely composed of Paludinæ.
Intermixed with these freshwater shells, there are
a few which seem to mark the occasional presence
of salt water, as for example, a species of Bulla,
together with an Oyster, and the Exogyra, a
genus of unimuscular bivalves allied to the oyster
(see Fig. 182.). The conclusion to be drawn from
the presence of a Corbula (see Fig. 183.) and

Fig. 182.   Fig. 183.

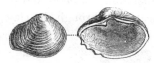

Exogyra bulla. Fitton.   *Corbula alata.* Fitton.
                                      Magnified.

Mytilus is more doubtful; for although these genera are for the most part marine, still there is a
Mytilus living in the Danube, and one species of
Corbula inhabits the river La Plata, in South
America, as well as the adjoining sea, while another is common to the Caspian, and the rivers
Don and Wolga. But admitting all these to have
been marine, they by no means outweigh the
evidence, both of a positive and negative kind, derived from shells in favour of the freshwater origin of the Wealden. In no part of this deposit
do we meet with ammonites, belemnites, terebratulæ, corals, sea-urchins, or other testacea and

zoophytes so characteristic of the chalk above, or the oolite below the Wealden.

Shells of the Cypris, an animal allied to the Crustacea, and before mentioned (p. 65.) as

Fig. 184.   Fig. 185.   Fig. 186.

Cypris
spinigera,
Fitton.

Cypris Valdensis, Fitton.
(C. faba, Min. Con. 485.)

Cypris tuberculata,
Fitton.

abounding in lakes and ponds, are also plentifully scattered through the clays of the Wealden, some-

Fig. 187.

times producing, like plates of mica, a thin lamination (see Fig. 187.). Similar cypriferous marls are found in the lacustrine tertiary beds of Auvergne, and in recent deposits of shell marl.

The fishes of the Wealden belong partly to the genera Pycnodus and Hybodus, forms common to the Wealden and Oolite (see Fig. 225.) ; but the teeth and scales of a species of Lepidotus are most widely diffused (see Fig. 188.). The general form of these fish was that of the carp tribe, although perfectly distinct in anatomical character, and more allied to the pike. The whole body was covered with large rhomboidal scales very thick, and having the exposed part covered with enamel. Most of

Fig. 188.

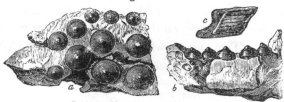

Lepidotus Mantelli, Agass. Wealden.

a. palate and teeth.   b. side view of teeth.   c. scale.

the species of this genus are supposed to have
been either river fish, or inhabitants of the coasts,
having not sufficient powers of swimming to advance into the deep sea.

Among the remains of vertebrata, those of reptiles form the most remarkable feature.   Some of
them belong to tortoises, such as the Trionyx and
Emys, genera now occurring in freshwater in
tropical regions.   Of Saurian lizards there are at
least five genera; the Crocodile, Plesiosaur, Me-
galosaur, Iguanodon, and Hylæosaur.   The Iguanodon, of which the remains were first discovered
by Mr. Mantell, was an herbivorous reptile, and
was regarded by Cuvier as more extraordinary
than any with which he was acquainted; for the
teeth, though bearing a great analogy to the
modern Iguanas which now frequent the tropical woods of America and the West Indies,
exhibit many striking and important differences
(see Fig. 190.).   It appears that they have been

worn by mastication; whereas the existing herbivorous reptiles clip and gnaw off the vegetable
productions on which they feed, but do not chew
them.   Their teeth, when worn, present an ap-

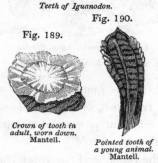

Teeth of Iguanodon.

Fig. 190.

Fig. 189.

Crown of tooth in
adult, worn down.
Mantell.

Pointed tooth of
a young animal.
Mantell.

pearance of having been chipped off, and never,
like the fossil teeth of the Iguanodon, have a flat
ground surface, (see Fig. 189.), resembling the
grinders of herbivorous mammalia.   From the
large bones, found in great numbers near these
teeth, and fairly presumed to belong to the same
animal, it is computed that the entire length of
this reptile could not have been less than seventy
feet.

The bones of birds of the order Grallæ or
waders have been discovered by Mr. Mantell in
the Wealden, and appear to be the oldest well-
authenticated examples of fossils of this class
hitherto found in Great Britain.*   But no portion

* Mantell, Proceedings Geol. Soc. vol. ii. p. 203.

of the skeleton of a mammiferous quadruped has yet been met with.

The vegetable remains, which are numerous, exhibit many characters of a tropical flora, some being allied to the living genera Cycas and Zamia (see Fig. 194.), others to large Equiseta. There are also Coniferæ allied to Araucaria, and other genera of warm climates (see Fig. 191), besides numerous ferns (see Fig. 192.).

Fig. 191.

Fig. 192.

Cone from the Isle of Purbeck, resembling the Dammara of the Moluccas. Fitton.

Sphenopteris gracilis (Fitton), *from near Tunbridge Wells.*
a. portion of the same magnified.

Passage of Wealden beneath Chalk. — It has been already seen that the chalk and green sand have an aggregate thickness of 1000 or sometimes 1500 feet. It is therefore a wonderful fact that after penetrating these rocks, we come down upon a subjacent *freshwater* formation from 800 to 1000 feet in thickness. The order of superposition is clear for we see the weald clay passing beneath the green-sand in various parts of Surrey, Kent, and Sussex ; and if we proceed from Sussex westward to the Vale

Fig. 193.

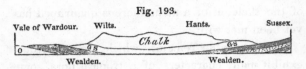

of Wardour, we there again observe the same
formation occupying the same relative position,
and resting on the oolite (see Fig. 193.).   Or if
we pass from the base of the south downs in Sussex, and cross to the Isle of Wight, we there
again meet with the same series reappearing beneath the green-sand, and we cannot doubt that
the beds are prolonged subterraneously, as indicated by the dotted lines in Fig. 194.

Fig. 194.

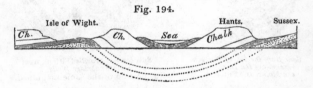

It has been already suggested that, during the
accumulation of the green-sand, there was a gradual
sinking down and submersion of land, by which
the wide open sea of the chalk was produced. But
the position of the Wealden points still more
forcibly to such a conclusion, and especially the
appearances exhibited at the point of junction of
the wealden, and the oolitic formation on which it
rests.   First, in regard to its junction with the
superincumbent lower green-sand, the beds of

this last, says Dr. Fitton, repose in the south-east of England, conformably upon those of the sub-jacent weald clay. There is no indication of disturbance : " To all appearance the change from the deposition of the freshwater remains to that of the marine shells, may have been effected simply by a tranquil submersion of the land to a greater depth beneath the surface of the waters." *

Portland dirt-bed and proofs of subsidence. — But when we examine the contact of the Purbeck beds, or *inferior* division of the wealden, with the Portland stone, or upper member of the oolite, some very singular phenomena are observed. Between the two formations, the marine and the freshwater, there intervenes in Portland a layer of dark matter, called by the quarrymen the " Dirt," or " Black dirt," which appears evidently to have been an ancient vegetable soil. It is from twelve to eighteen inches thick, is of a dark brown or black colour, and contains a large proportion of earthy lignite. Through it are dispersed rounded fragments of stone, from three to nine inches in diameter, in such numbers that it almost deserves the name of gravel. Many silicified trunks of coniferous trees, and the remains of plants allied to the Zamia and Cycas are buried in this dirt-bed (see figure of living Zamia).

* Geol. of Hastings, p. 28

Fig. 195.

Zamia spiralis ; Southern Australia. *

These plants must have become fossil on the
spots where they grew.   The stumps of the trees
stand erect for a height of from one to three feet,
and even in one instance to six feet, with their
roots attached to the soil at about the same distances from one another as the trees in a modern
forest.†   The carbonaceous matter is most abundant immediately around the stumps, and round
the remains of fossil *Cycadeæ.* ‡

Besides the upright stumps above mentioned,
the dirt-bed contains the stems of silicified trees
laid prostrate.   These are partly sunk into the

* See Flinder's Voyage.

† Mr. Webster first noticed the erect position of the trees
and described the Dirt-bed.   The account here given is
drawn from Dr. Buckland and Mr. De la Beche, Geol.
Trans., Second Series, vol. iv. p. 1.; Mantell, Geol. of S. E.
of England, p. 336.; and Dr. Fitton, Geol. Trans., Second
Series, vol. iv. p. 220.

‡ Fitton, ibid. pp. 220, 221.

black earth, and partly enveloped by a calcareo-
siliceous slate which covers the dirt-bed.   The
fragments of the prostrate trees are rarely more
than three or four feet in length; but by joining
many of them together, trunks have been restored
having a length from the root to the branches of
from 20 to 23 feet, the stems being undivided
for 17 or 20 feet, and then forked.   The diameter of these near the roots is about one foot.*
Root-shaped cavities were observed by Professor
Henslow to descend from the bottom of the dirt-
bed into the subjacent Portland stone, so that the
uppermost beds of the Portland limestone, though
now solid, were in a soft and penetrable state
when the trees grew.†

The thin layers of calcareous slate (Fig. 196.),

Fig. 196.

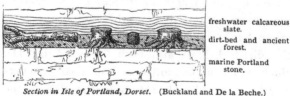

freshwater calcareous
slate.

dirt-bed and ancient
forest.

marine Portland
stone.

Section in Isle of Portland, Dorset.   (Buckland and De la Beche.)

were evidently deposited tranquilly, and would
have been horizontal but for the protrusion of the

* Fitton, Geol. Trans., Second Series, vol. iv. pp. 220, 221.
† Buckland and De la Beche, Geol. Trans., Second
Series, vol. iv. p. 16.

stumps of the trees, around the top of each of which they form hemispherical concretions.

The dirt-bed is by no means confined to the island of Portland, but is seen in the same relative position in a cliff east of Lulworth Cove, in Dorsetshire, where, as the strata have been disturbed, and are now inclined at an angle of 45°, the stumps of the trees are also inclined at the

Fig. 197.

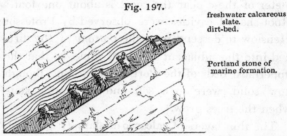

freshwater calcareous slate.
dirt-bed.

Portland stone of marine formation.

Section in cliff east of Lulworth Cove.   (Buckland and De la Beche.)

same angle in an opposite direction — a beautiful illustration of a change in the position of beds originally horizontal (see Fig. 197.). Traces of the dirt-bed have also been observed by Dr. Buckland, about two miles north of Thame, in Oxfordshire; and by Dr. Fitton, in the cliffs of the Boulonnois, on the French coast: but, as might be expected, this freshwater deposit is of limited extent when compared to most marine formations.

From the facts above described, we may infer, first, that the superior beds of the oolite, which are full of marine shells, became dry land, and covered by a forest, throughout a portion of the space now occupied by the south of England, the climate being such as to admit the growth of the zamia and cycas. 2dly. This land at length sank down and was submerged with its forests beneath a body of freshwater, from which sediment enveloping fluviatile shells was deposited. 3dly. " The regular and uniform preservation of this thin bed of black earth over a distance of many miles, shows that the change from dry land to the state of a freshwater lake or estuary, was not accompanied by any violent denudation, or rush of water, since the loose black earth, together with the trees which lay prostrate on its surface, must inevitably have been swept away had any such violent catastrophe then taken place." *

The dirt-bed has been described above in its most simple form, but in some sections the appearances are more complicated. The forest of the dirt-bed was not everywhere the first vegetation which grew in this region. Two other beds of carbonaceous clay, one of them containing *Cycadeæ* in an upright position have been found below it †, which implies other oscillations in the level of the same ground, and its alternate occupation by land and water more than once. There

* Buckland and De la Beche, Geol. Trans., Second Series, vol. iv. p. 16.

† Fitton, Geol. Trans., Second Series, vol. iv. p. 223.

must have been, first, the sea in which the corals
and shells of the oolite grew; then, land, which supported a vegetable soil with Cycadeæ; then, a lake
or estuary, in which freshwater strata were deposited; then, again, land, on which other Cycadeæ
and a forest of dicotyledonous trees flourished;
then, a second submergence under freshwater, in
which the wealden strata were gradually formed;
and, finally, in the cretaceous period, a return over
the same space of the ocean.

To imagine such a series of events will appear
extravagant and visionary to some who are not
aware that similar changes occur in the ordinary
course of nature; and that large areas near the
sea are now subject to be laid dry, and then submerged, after remaining years covered with houses
and trees.*

In some of these modern revolutions, such as
have been witnessed in the delta of the Indus,
in Cutch, we have instances of land being permanently laid under the waters, both of the river
and the sea, without the soil and its shrubs being
swept away; but such preservation of an ancient
soil must be a rare exception to the general rule,
for it would be destroyed by denuding waves and
currents, unless the land sank suddenly down to

* For an account of recent movements of land attended
by such consequences, see Principles of Geology, Index,
" Cutch," " Sindree," &c.

a great depth, or unless its form was such as to exclude the free ingress of the sea.

Notwithstanding the enormous thickness of the wealden, exceeding in some places perhaps 1000 feet, there are many grounds for believing that the whole of it was a deposit in water of moderate depth, and often extremely shallow. This idea may seem startling at first, yet such would be the natural consequence of a gradual and continuous sinking of the ground in an estuary or bay, into which a great river discharged its turbid waters. By each foot of subsidence, the fundamental rock, such as the Portland oolite, would be depressed one foot farther from the surface of the ocean; but the bay would not be deepened, if new strata of mud and sand should raise the bottom one foot. On the contrary, such sand and mud might be frequently laid dry at low water, or overgrown for a season by a vegetation proper to marshes. At different heights in the Hastings Sand in the middle of the Wealden, we find again and again slabs of sandstone with a strong ripple-mark, and between these slabs beds of clay many yards thick. In some places, as at Stammerham, near Horsham, there are indications of this clay having been exposed so as to dry and crack before the next layer was thrown down upon it. The open cracks in the clay have served as moulds, of which casts have been taken in relief, and which are, there-

fore, seen on the lower surface of the sandstone *
(see Fig. 198.).

Fig. 198.

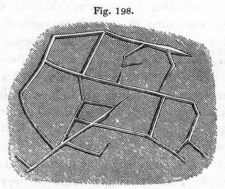

Underside of slab of sandstone about one yard in diameter ;
Stammerham, Sussex.

Near the same place a reddish sandstone occurs
in which are innumerable traces of a fossil vegetable, apparently Sphenopteris, the stems and
branches of which are disposed as if the plants
were standing erect on the spot where they originally grew, the sand having been gently deposited upon and around them ; and similar
appearances have been remarked in other places
in this formation. †   In the same division also of
the wealden, at Cuckfield, is a bed of gravel or
conglomerate, consisting of water-worn pebbles of
quartz and jasper, with rolled bones of reptiles.

* Observed by Mr. Mantell and myself in 1831.
† Mantell, Geol. of S. E. of England, p. 244.

These must have been drifted by a current, probably in water of no great depth.

The occasional presence of oysters in the Purbeck limestone, and throughout the Hastings sand and Weald clay, proves that the waters of the sea sometimes found access into the estuary *, whether in consequence of subsidence, or in seasons when the body of freshwater was lessened in volume.

Geographical extent. — The Wealden strata have been traced about 200 English miles from west to east, from Lulworth Cove to near Boulogne, in France, and about 220 miles from north-west to south-east, from Whitchurch, in Buckinghamshire, to Beauvais, in France. If the formation be continuous throughout this space, which is very doubtful, it does not follow that the whole was contemporaneous; because in all likelihood the physical geography of the region underwent frequent change throughout the whole period, and the estuary may have altered its form, and even shifted its place. Yet some modern deltas are of vast size, as for example, that of the newly discovered Quorra, or Niger, in Africa, which stretches into the interior for more than 170 miles, and occupies, it is supposed, a space of more than 300 miles along the coast; thus forming a surface of

* Fitton, Geol. Trans., 2d Ser., vol. iv. p. 321.

more than 25,000 square miles, or equal to about
one half of England.*

I have stated that the Wealden has been observed near Beauvais, in France; and the locality
is marked in the section at p. 315. It is called
" the country of Bray;" and resembles in structure the English Weald between the north and
south downs. In a similar manner the green-
sand crops out from beneath the chalk, and freshwater strata from beneath the green-sand. One
member of the series, a fine whitish sand, contains
impressions of ferns, considered by M. Adolphe
Brongniart as identical with *Lonchopteris Mantelli*,
a plant found frequently in the Wealden. I examined part of the valley of Bray in company
with M. Graves, in 1833, and I observed that
the sand last mentioned, with its vegetable remains, was intercalated between two sets of marine
strata, containing trigoniæ, and referred by French
geologists to the lower green-sand. In the same
country of Bray, and associated with the same
formation, is a limestone resembling the Purbeck
marble, and containing a Paludina which seems
specifically identical with that of Purbeck.

If it be asked where the continent was placed
from the ruins of which the Wealden strata were

* Fitton, Geol. of Hastings, p. 58.; who cites Lander's
Travels.

derived, and by the drainage of which a great river was fed, we are half tempted to speculate on the former existence of the Atlantis of Plato. The story of the submergence of an ancient continent, however fabulous in history, may be true as a geological event. Its disappearance may have been gradual; and we need not suppose that the rate of subsidence was hastened at the period when the displacement of a great body of freshwater by the cretaceous sea took place. Suppose the mean height of the land drained by the river of the Wealden estuary to have been no more than 800 or 1000 feet; in that case, all except the tops of the mountains would be covered as soon as the fundamental oolite and the dirt-bed were sunk down about 1000 feet below the level which they occupied when the forest before-mentioned was growing. Towards the close of the period of this subsidence, both the sea would encroach and the river diminish in volume more rapidly; yet in such a manner, that we may easily conceive the sediment at first washed into the advancing sea to have resembled that previously deposited by the river in the estuary. In fact, the upper beds of the Wealden, and the inferior strata of the lower green-sand, are not only conformable, but of similar mineral composition.

It is also a remarkable fact, that the same *Iguanodon Mantelli* which is so conspicuous a fossil in the

Wealden, has recently been discovered near Maidstone, in the overlying Kentish rag, or marine limestone of the lower green-sand. Hence we may infer that some of the saurians which inhabited the country of the great river, continued to live when part of the country had become submerged beneath the sea. Thus, in our own times, we may suppose the bones of large alligators to be frequently entombed in recent freshwater strata in the delta of the Ganges. But if part of that delta should sink down so as to be covered by the sea, marine formations might begin to accumulate in the same space where freshwater beds had previously been formed; and yet the Ganges might still pour down its turbid waters in the same direction, and carry the carcasses of the same species of alligator to the sea, in which case their bones might be included in marine as well as in subjacent freshwater strata.

Age of the Wealden. — Some geologists have classed the Wealden as a member of the cretaceous group, while others have considered it as more nearly connected with the antecedent oolitic deposits; nor is it easy to decide which opinion is preferable, because the organic remains of the cretaceous and oolitic groups are marine, while those of the interposed Wealden are almost all freshwater. The testacea and plants of the latter appear as yet to be specifically distinct from those of any other formation; but if we examine the reptiles, it appears that the *Megalosaurus Bucklandi* is common to the Oolite and Wealden, the teeth and bones of this great saurian occurring both in the limestone of Stonesfield and in the Hastings sand.

There are also some *generic* forms, both of reptiles and fish, common to the Oolite and Wealden, and not yet discovered in the Chalk. Vertebræ, for example, of the Plesiosaurus are not confined to the oolite and lias, but have been also found in the Wealden; and the Lepidotus, a genus of fish very characteristic of the Wealden, is unknown in the cretaceous group, while it is abundant in the oolitic series.

On the other hand, the same species of Iguanodon has been already mentioned as decidedly common to the Wealden and green-sand.

In Scotland, and in different parts of the Continent, marine deposits have been found which are supposed to have been coeval with the Wealden, and which are intermediate in fossil characters as in position between the Cretaceous and Oolitic systems.* They may have been contemporaneous deltas of other rivers flowing from the same ancient continent.

Absence of mammalia. — Among the numerous

* See Fitton, Geol. Trans., Second Series, vol. iv. p. 328., and his references.

fossils of the Wealden, no remains of mammalia have been hitherto detected; whereas we should naturally expect, on examining the deposits recently formed at the mouths of the Quorra, Indus, or Ganges, to find, not only the bones of birds and of amphibious and land reptiles, but also those of such warm-blooded quadrupeds as frequent the banks of rivers, or, like the hippopotamus, inhabit their waters. Would not the same current of water which drifted down and rolled the bones of the lizards, tortoises, and fish of the Wealden, have also swept down into the delta some fragments at least of mammiferous bones, had any animals of the highest class been then in existence? As a general rule, indeed, we cannot lay much stress on mere negative evidence; and it may be well to notice, that although so many teeth of the Iguanodon have been collected, it is only of late that a single small portion of a jaw of one of these gigantic lizards was obtained. Perhaps, in like manner, some bone or tooth of a fossil quadruped will one day be found. We may at least say, that we have at present no example of a continent covered with a luxuriant vegetation, and forests inhabited by large saurians, both aquatic and terrestrial, and by birds, yet at the same time entirely destitute of warm-blooded quadrupeds. The nearest analogy to this state of things is that of New Zealand; and this fact

will be more particularly alluded to in the sequel.
(See p. 442.)

In conclusion I may remark, that from the
time of the commencement of the Wealden, to far
on in the Cretaceous period, we have signs of subsidence, and consequent diminution of land. But
after the chalk was formed, or during the tertiary
periods, we have, on the contrary, proofs of an increase of land in Europe. But we must not extend these generalizations to the whole surface of
the globe; for other large areas may have been
growing more and more continental during the
cretaceous, and more and more oceanic during
the tertiary periods, the direction of the prevailing subterranean movement being reversed.

CHAPTER XVII.

OOLITE AND LIAS.

Subdivisions of the Oolitic group — Fossil shells — Corals
in the calcareous divisions only — Buried forest of Encrinites in Bradford clay — Changes in organic life during
accumulation of Oolites — Characteristic fossils — Signs
of neighbouring land and shoals — Supposed cetacea in
Oolite. — Oolite of Yorkshire and Scotland.

OOLITE. — Below the freshwater group last described, or, where this is wanting, immediately
beneath the Cretaceous formation, a great series
of marine strata, commonly called " the Oolite," occurs in many parts of Europe. This group has been
so named, because, in England and other places
where it was first examined, the limestones belonging to it had an oolitic structure (see p. 29.).
These rocks occupy in England a zone which is
nearly thirty miles in average breadth, and extends across the island, from Yorkshire on the
north-east, to Dorsetshire on the south-west.*
Their mineral characters are not uniform throughout this region; but the following are the names

* For details respecting this formation in England, see
Conybeare and Phillips's Geology, chap. iii.

of the principal subdivisions observed in the central and south-eastern parts of England: —

OOLITE.

Upper $\begin{cases} a. & \text{Portland stone and sand.} \\ b. & \text{Kimmeridge clay.} \end{cases}$

Middle $\begin{cases} c. & \text{Coral rag.} \\ d. & \text{Oxford clay.} \end{cases}$

Lower $\begin{cases} e. & \text{Cornbrash and Forest Marble.} \\ f. & \text{Great Oolite and base of Fullers' earth.} \\ g. & \text{Inferior Oolite.} \end{cases}$

The Lias then succeeds to the Inferior Oolite.

The upper oolitic system of the above Table has usually the Kimmeridge clay for its base, and the middle oolitic system the Oxford clay. The lower system reposes on the Lias, an argillo-calcareous formation, which some include in the lower oolite, but which will be treated of separately in the next chapter. Many of these subdivisions are distinguished by peculiar organic remains; and though varying in thickness, may be traced in certain directions for great distances, especially if we compare the part of England to which the above-mentioned type refers with the north-west of France, and the Jura mountains, which separate that country from Switzerland, and in which, though distant above 400 geographical miles, the analogy to the English type above mentioned is more perfect than in Yorkshire or Normandy.

To enter upon a systematic description of this complicated series of strata would require many chapters; the following facts, therefore, are selected from a multitude of others, with a view of illustrating the origin of the oolitic rocks, and of showing the state of organic life and geographical condition of part of the globe when they were formed.

In almost all the minor divisions enumerated in the above Table, Ammonites and Belemnites are found (see Figs. 213. 215.), but of species different from those of the cretaceous period. The ammonites are of various sizes, from the size of a small carriage-wheel to less than an inch diameter.

It is not uncommon to find belemnites in different members of the series, with full grown serpulæ attached to them. As these shells, like the bone of the cuttle-fish, so often thrown on our shores, were internal, it is clear, that after the death of the cephalopod the belemnite remained for some time unburied at the bottom of the sea, so that the serpulæ grew upon it.

These cephalopoda, swimming about in the open sea, left their shells to be imbedded indifferently in whatever sediment was then in the course of deposition, whether calcareous or argillaceous. But the corals are almost entirely confined to the limestones, and are wanting in the dense formations of interposed clay, as also in the lias, these zoophytes requiring, not only carbonate of lime for their support, and clear water, but a bottom remaining for years unchanged, either by the shifting of sand or the accession of fresh sediment.

In the Upper Oolite of England, corals are rare, although one species is found plentifully at Tisbury, in Wiltshire, in the Portland sand, converted into flint and chert, the original calcareous matter being replaced by silex. (Fig. 199.) One of the limestones of the Middle Oolite has been called the " Coral Rag," because it consists, in part, of continuous beds of petrified corals, for the most part retaining the position in which they grew at the bottom of the sea. They belong chiefly to the genera Caryophyllia (Fig. 200.), Agaricia, and

Fig. 199.

Fig. 200.

Columnaria oblonga, Blainv.
Upper Oolite, Tisbury.

Caryophyllia annularis, Parkin.
Coral rag, Steeple Ashton.

Astrea, and sometimes form masses of coral fifteen feet thick. These coralline strata extend through the calcareous hills of the N. W. of Berkshire, and north of Wilts, and again recur in Yorkshire,

near Scarborough. Although the name of coral
rag has been thus appropriated, there are portions
of the lower oolite, as for example the Great and
Inferior Oolite (*f. g.* Table, p. 369.), which are
equally entitled in many places to be called coralline limestones. Thus the Great Oolite near
Bath contains various corals, among which the
Eunomia radiata (Fig. 201.) is very conspicuous,

Fig. 201.

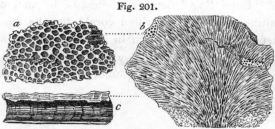

Eunomia radiata, Lamouroux.

a. section transverse to the tubes.
b. vertical section, showing the radiation of the tubes.
c. portion of interior of tubes magnified, showing striated surface.

single individuals forming masses several feet in
diameter; and having probably required, like the
large existing brain-coral (*Meandrina*) of the tropics, many centuries before their growth was
completed.

Different species of Crinoideans, or stone-lilies,
are also common in the same rocks with corals;
and, like them, must have enjoyed a firm bottom,
where their root, or base of attachment, remained
undisturbed for years (*c* Fig. 202.) Such fossils,

Fig. 202.

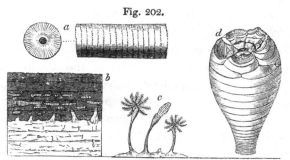

Apiocrinites rotundus, or Pear Encrinite; Miller. Fossil at Bradford, Wilts.

a. Stem of Apiocrinites, and one of the articulations, natural size.
b. Section at Bradford of great oolite and overlying clay, containing the fossil encrinites. See text.
c. Three perfect individuals of the Apiocrinite, represented as
they grew on the surface of the Great Oolite.
d. Body of the *Apiocrinites rotundus.*

therefore, are almost confined to the limestones;
but an exception occurs at Bradford, near Bath,
where they are enveloped in clay. In this case,
however, it appears that the solid upper surface
of the " Great Oolite " had supported, for a time,
a thick submarine forest of these beautiful zoophytes, until the clear and still water was invaded
by a current charged with mud, which threw
down the stone-lilies, and broke most of their
stems short off near the point of attachment. The
stumps still remain in their original position; but
the numerous articulations once composing the
stem, arms, and body of the zoophyte, were scattered at random through the argillaceous deposit
in which some of them now lie prostrate. These

appearances are represented in the section *b*, Fig.
202., where the darker strata represent the Bradford clay, a member of the Forest marble (*e.* Table,
p. 369.). The upper surface of the calcareous
stone below is completely incrusted over with a
continuous pavement, formed by the stony roots
or attachments of the Crinoidea; and besides this
evidence of the length of time they had lived on
the spot, we find great numbers of single vertebræ, or circular plates of the stem and body of
the encrinite, covered over with serpulæ. Now
these serpulæ could only have begun to grow after
the death of some of the stone-lilies, parts of
whose skeletons had been strewed over the floor
of the ocean before the irruption of argillaceous
mud. In some instances we find that, after the
parasitic serpulæ were full grown, they had be-

Fig. 203.

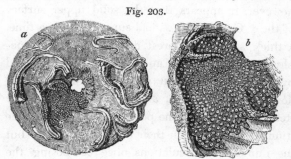

a. Single vertebra, or articulation of an Encrinite overgrown with
  serpulæ and corals. Natural size. Bradford clay.
b. Portion of the same magnified, showing the coral *Berenicea
  diluviana* covering one of the serpulæ.

come incrusted over with a coral, called *Berenicea diluviana ;* and many generations of these polyps had succeeded each other in the pure water before they became fossil.

We may, therefore, perceive distinctly that, as the pines and cycadeous plants of the ancient Portland Forest were killed by submergence under fresh water, and soon buried under muddy sediment, so an invasion of argillaceous matter put a sudden stop to the growth of the Bradford Encrinites, and led to their preservation in marine strata.*

Such differences in the fossils as distinguish the calcareous and argillaceous deposits from each other, would be described by naturalists as arising out of a difference in the *stations* of species; but besides these, there are variations in the fossils of the higher, middle, and lower part of the oolitic series, which must be ascribed to that great law of change in organic life by which distinct assemblages of species have been adapted, at successive geological periods, to the varying conditions of the habitable surface. In a single district it is difficult to decide how far the limitation of species to certain minor formations has been due to the local influence of *stations*, or how far it has been caused by time, or the creative and destroying

* For a fuller account of these Encrinites, see Buckland's Bridgewater Treatise, vol. i. p. 429.

law above alluded to. But we recognize the
reality of the last mentioned influence, when
we contrast the whole oolitic series of England
with that of parts of the Jura, Alps, and other
distant regions, where there is scarcely any lithological resemblance; and yet some of the same
fossils remain peculiar in each country to the
Upper, Middle, and Lower Oolite formations respectively. Mr. Thurmann has shown how remarkably this fact holds true in the Bernese Jura,
although the argillaceous divisions, so conspicuous
in England, are feebly represented, and some entirely wanting.

Amongst the characteristic fossils of the Upper
Oolite, may be mentioned the *Ostrea deltoidea*
(Fig. 205.), found in the Kimmeridge clay through-
out England and the north of France, and also in
Scotland, near Brora. The *Gryphæa virgula* (Fig.

FOSSILS OF THE OOLITE.

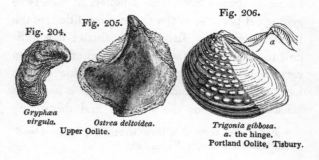

Fig. 206.

Fig. 205.

Fig. 204.

*Gryphæa
virgula.*

Ostrea deltoidea.
Upper Oolite.

a

Trigonia gibbosa.
a. the hinge.
Portland Oolite, Tisbury.

Fig. 207.   Fig. 208.

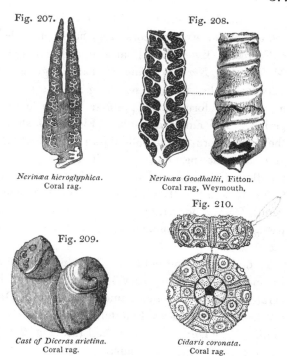

Nerinæa hieroglyphica.   Nerinæa Goodhallii, Fitton.
Coral rag.   Coral rag, Weymouth.

Fig. 210.

Fig. 209.

Cast of Diceras arietina.   Cidaris coronata.
Coral rag.   Coral rag.

204.), also met with in the same clay near Oxford, and so abundant in the upper oolite of parts
of France as to have caused the deposit to be
termed " marnes à grypheés virgule."   Near
Clermont, in Argonne, a few leagues from St.
Menehould, these indurated marls crop out from
beneath the gault; and, on decomposing, leave
the surface of every ploughed field literally strewed
over with fossil oysters.

One of the limestones of the Jura, referred to
the age of the English coral rag, has been called
" Nerinæan limestone" (Calcaire à Nérinées) by
M. Thirria; *Nerinæa* being an extinct genus of
univalve shells, much resembling the Cerithium
in external form, and peculiar to the oolitic
period.   The annexed section (Fig. 207.) shows
the curious form of the hollow part of each whorl,
and also the perforation which passes up the
middle of the columella.   *N. Goodhallii* (Fig.
208.) is another English species of the same
genus, from a formation which seems to form a
passage from the Kimmeridge clay to the coral
rag.*

A division of the oolite in the Alps, regarded
by most geologists as coeval with the English
coral rag, has been often named " Calcaire à Di-
cerates," or " Diceras limestone," from its containing abundantly a bivalve shell (see Fig. 209.)
of a genus allied to the Chama.

Among the characteristic shells of the Inferior
Oolite, I may instance *Terebratula spinosa* (Fig.
211.), *Pholadomya fidicula* (Fig. 212.), *Belemnites
hastatus* (Fig. 213.), and *Terebratula digona* (Fig.
216.)

As illustrations of shells having a great vertical
range, I may allude to *Trigonia gibbosa* (Fig.

* Fitton, Geol. Trans., Second Series, vol. iv. pl. 23. fig. 12.

Fig. 211.   Fig. 212.

Terebratula spinosa.   *a. Pholadomya fidicula.* Inferior Oolite.
Inferior Oolite.   *b. Heart-shaped anterior termination of the same.*

Fig. 213.

Belemnites hastatus.   Inferior Oolite.

Fig. 214.   Fig. 215.

Orbicula reflexa, Sow.   *Ammonites striatulus*, Sow.
a. upper valve.   Inferior Oolite and Lias.
b. lower or attached valve, and
   showing part of the upper.

Fig. 216.   Fig. 217.

Terebratula digona.   *Ostrea Marshii.*
Inferior Oolite.   Middle and Lower Oolite.

205.), which abounds in the Portland stone of
Wiltshire, and the Inferior Oolite of Yorkshire.*
Also *Ostrea Marshii* (Fig. 217.), common to the

* See Williamson, Proceedings Geol. Soc. No. 47.

Cornbrash of Wilts and the Inferior Oolite of Yorkshire; and, lastly, *Orbicula reflexa* (Fig. 214.) and *Ammonites striatulus* (Fig. 215.), fossils common to the Inferior Oolite and Lias.

Such facts by no means invalidate the general rule, that certain fossils are good chronological tests of geological periods; but they serve to caution us against attaching too much importance to single species, some of which may have a wider, others a more confined vertical range. We have before seen that, in some of the tertiary formations, some species occur both in the older and newer groups, yet these groups may be distinguishable from one another by a comparison of the whole assemblage of fossil shells proper to each.

Signs of neighbouring land and shoals. — The corals and shells above alluded to, and the fish, crustacea, and other accompanying fossils, sufficiently attest the marine origin of the oolitic strata in general. Yet there are frequent signs of shallow water and of neighbouring land; and these are the more worthy of attention, as they by no means diminish as we proceed downwards to the inferior parts of the oolitic series. Had the bottom of the sea in Europe been unmoved during the entire oolitic period, the first, or oldest beds of the oolite, must have been accumulated in the deepest water, the middle oolite in water of less

depth, and the upper in the shallowest of all.
The appearances about to be described militate
against this conclusion. The Kimmeridge clay,
in the Upper Oolite, consists, in great part, of a
bituminous shale, sometimes forming an impure
coal several hundred feet in thickness. In some
places in Wiltshire it much resembles peat; and
the bituminous matter may have been, in part at
least, derived from the decomposition of vegetables. But as impressions of plants are rare in
these shales, which contain ammonites, oysters,
and other marine shells, the bitumen may perhaps
be of animal origin. The occurrence, however, of
fossil wood in the Upper Oolite shows that there
were then lands from which plants were drifted
into the sea.

The celebrated lithographic stone of Solenhofen, in Bavaria, belongs to one of the upper
divisions of the oolite, and affords a remarkable
example of the variety of fossils which may be
preserved under favourable circumstances, and
what delicate impressions of the tender parts of
certain animals and plants may be retained where
the sediment is of extreme fineness. Although
the number of testacea in this slate is small, and
the plants few, and those all marine, Count Munster had determined no less than 237 species of
fossils when I saw his collection in 1833; and
among them no less than seven *species* of flying

lizards, or pterodactyls, six saurians, three tortoises, sixty species of fish, forty-six of crustacea, and twenty-six of insects. These insects, among which is a libellula, or dragon-fly, must have been blown out to sea, probably from the same land to which the flying lizards, and other contemporaneous reptiles, resorted.

In one of the upper members of the *Inferior Oolite* of England the ripple-mark is distinctly seen throughout a considerable thickness of thin fissile beds of a coarsely oolitic limestone. The rippled slabs are used for roofing, and have been traced over a broad band of country from Bradford, in Wilts, to Tetbury in Gloucestershire. These calcareous slabs, or tile-stones, are separated from each other by thin seams of clay, which have been deposited upon them, and have taken their form, preserving the undulating ridges and furrows of the sand in such complete integrity, that the impressions of small footsteps, apparently of crabs, which walked over the soft wet sands, are still visible. In the same stone the claws of crabs, fragments of echini, broken shells, pieces of drift wood, and other signs of a neighbouring beach, are observed.

The slate of Stonesfield has lately been shown by Mr. Lonsdale to lie at the base of the Inferior Oolite. It is an oolitic shelly limestone, only six feet thick, but very rich in organic remains. It contains some pebbles of a rock very similar to itself, and with them the fossil remains of belemnites, trigoniæ, and other marine shells. Besides fragments of wood, which occur in all parts of the oolitic group, there are many impressions of ferns, cycadeæ, and other terrestrial plants. Several insects also, and among the rest, the wing-covers of beetles, are perfectly preserved (see Fig. 218.), some of them approaching nearly to the genus Buprestis.* The remains, also, of many genera of reptiles, such as Plesiosaurus, Crocodile, and Pterodactyl, have been discovered in the same limestone ; and, what is still more remarkable, the jaws of at least two species of mammiferous quadrupeds, allied to the Didelphys, or opossum. These fossils afford the only example yet known of terrestrial mammalia in rocks of a date anterior to the Eocene period.

Fig. 218.

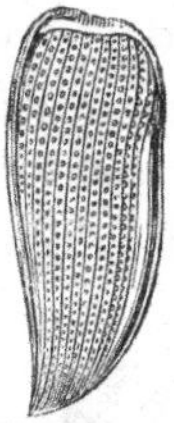

Elytron of Buprestis ; Stonesfield.

This exception is the more deserving of notice, because even no cetacea have as yet been observed in any secondary strata, although certain bones, from the great oolite of Enstone, near Woodstock, in Oxfordshire, have been cited, on the authority of Cuvier, as referable to this class. Dr. Buckland, who has stated this in his late Bridgewater Trea-

* See Buckland's Bridgewater Treatise.

tise *, has had the kindness to send me the supposed ulna of a whale, in order that Mr. Owen might examine into its claims to be considered as cetaceous. It is the opinion of that eminent comparative anatomist, that it cannot have belonged to the cetacea, because the fore-arm in these marine mammalia is invariably much flatter, and devoid of all muscular depressions and ridges, one of which is so prominent in the middle of this bone (see Fig. 219.). In saurians, on the contrary, such ridges

Fig. 219.

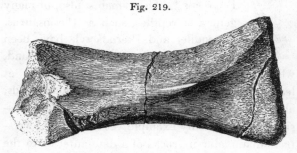

Bone of a reptile, formerly supposed to be the ulna of a Cetacean ; from the Oolite of Enstone, near Woodstock.

exist for the attachment of muscles; and to some animal of that class the bone is probably referable.

Oolite of Yorkshire and Scotland. — North of the Humber, in Yorkshire, the Inferior Oolite assumes a form very different from that which distinguishes it in the south. It may there be called a coal formation, as it contains much vegetable matter,

* Vol. i. p. 115.

and coal, interstratified with sand and sandstones. The high state of preservation and number of the plants render it probable that land was not far distant. The same may be said of the oolitic coal of Brora, on the south-east coast of Sutherlandshire, in Scotland, where the Inferior Oolite contains coal, one bed of which is $3\frac{1}{2}$ feet in thickness. The plants resemble those in the Yorkshire oolite, and a great number of the associated marine shells and other fossils are the same * ; but the mineral characters of the sandstone, shale, and calcareous grit, differ considerably.

* Murchison, Geol. Trans., vol. ii. Second Series.

CHAPTER XVIII.

Mineral character of Lias — Name of Gryphite limestone — Fossil fish — Ichthyodorulites — Reptiles of the Lias — Ichthyosaur and Plesiosaur — Newly-discovered marine Reptile of the Galapagos Islands — Sudden death and burial of fossil animals in Lias — Origin of the Oolite and Lias, and of alternating calcareous and argillaceous formations — Physical geography — Vales of clay — Hills and escarpments of limestone.

LIAS. The English provincial name of Lias has been very generally adopted for a formation of argillaceous limestone, marl, and clay, which forms the base of the oolite, and is classed by many geologists as part of that group. They pass, indeed, into each other in some places, as near Bath, a sandy marl called the marlstone of the Lias being interposed, and partaking of the mineral characters of the upper lias and inferior oolite. These last mentioned divisions have also some

Fig. 220.

fossils in common, such as the *Avicula inæquivalvis* (Fig. 220.). Nevertheless the Lias may be traced throughout a great part of Europe as a separate and

Avicula inæquivalvis, Sow.

independent group, of considerable thickness, varying from 500 to 1000 feet, containing many peculiar fossils, and having a very uniform lithological aspect. Although usually conformable to the oolite, it is sometimes, as in the Jura, unconformable. Thus, in the environs of Lons-le-Saulnier, for instance, the strata of lias are inclined at an angle of about 45°, while the incumbent oolitic marls are horizontal.

The peculiar aspect which is most characteristic of the Lias in England, France, and Germany, is an alternation of thin beds of limestone, with a light brown weathered surface, separated by dark-coloured narrow argillaceous partings, so that the quarries of this rock, at a distance, assume a striped and riband-like appearance.*

Although the prevailing colour of the limestone of this formation is blue, yet some beds of the lower lias are of a yellowish white colour, and have been called white lias. In some parts of France, near the Vosges mountains, and in Luxembourg, M. E. de Beaumont has shown that the lias containing *Gryphæa arcuata*, *Plagiostoma giganteum*, and other characteristic fossils, becomes arenaceous; and around the Hartz, in Westphalia and Bavaria, the inferior parts of the lias are sandy, and sometimes afford a building stone called by the Germans quadersandstein.

* Conyb. and Phil. p. 261.

s 2

The name of Gryphite limestone has sometimes
been applied to the lias, in consequence of the
great number of shells which it contains of a
species of oyster, or Gryphæa (Fig. 221.). Many
cephalopoda, also, such as Ammonite, Belemnite,
and Nautilus (Fig. 222.), prove the marine origin
of the formation.

Fig. 221.   Fig. 222.

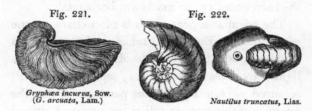

Gryphæa incurva, Sow.
(G. arcuata, Lam.)   Nautilus truncatus, Lias.

The fossil fish resemble generically those of the
oolite, belonging all, according to M. Agassiz, to
extinct genera, and differing remarkably from the
ichthyolites of the cretaceous period. Among
them is a species of Lepidotus (L. gigas, Ag.) (Fig.
223.), which is found in the lias of England,

a   Fig. 223.

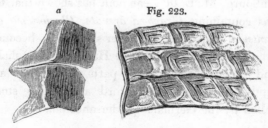

Scales of Lepidotus gigas, Agas.
a. two of the scales detached

France, and Germany.*   This genus was before
mentioned (p. 348.) as occurring in the Wealden,
and is supposed to have frequented both rivers
and coasts.   The teeth of a species of Acrodus,
also, are very abundant in the lias (Fig. 224.)

Fig. 224.

Acrodus nobilis, Agas. tooth ; commonly called fossil leach.
Lias, Lyme Regis, and Germany.

But the remains of fish which have excited
more attention than any others, are those large
bony spines called ichthyodorulites (*a.* Fig. 225.),

Fig. 225.

Hybodus reticulatus, Agas. Lias, Lyme Regis.
a. Part of fin, commonly called Ichthyodorulite.
b. Tooth.

which were once supposed by some naturalists to
be jaws, and by others weapons, resembling those
of the living Balistes and Silurus; but which M.
Agassiz has shown to be neither the one nor the
other.   The spines, in the genera last mentioned,

* Agassiz, Pois. Fos., vol. ii. tab. 28, 29.

articulate with the backbone, whereas there are
no signs of any such articulation in the ichthyodorulites. These last appear to have been bony
spines which formed the anterior part of the
dorsal fin, like that of the living genera Cestracion
and Chimæra (see *a*. Fig. 226.). In both of these

Fig. 226.

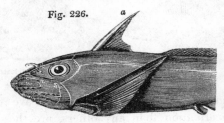

Chimæra monstrosa. *
a. Spine forming anterior part of the dorsal fin.

genera, the posterior concave face is armed with
small spines like that of the fossil Hybodus (Fig.
225.), one of the shark family found fossil at
Lyme Regis. Such spines are simply imbedded
in the flesh, and attached to strong muscles.
" They serve," says Dr. Buckland, " as in the Chimæra (Fig. 226.), to raise and depress the fin,
their action resembling that of a moveable mast,
raising and lowering backwards the sail of a
barge."†

Reptiles of the Lias. — It is not, however, the
fossil fish which form the most striking feature in

* Agassiz, Poissons Fossiles, vol. iii. tab. C. fig. 1.
† Bridgewater Treatise, p. 290.

the organic remains of the Lias; but the reptiles, which are extraordinary for their number, size, and structure.   Among the most singular of these are several species of Ichthyosaurus and Plesio- saurus.   The genus Ichthyosaurus, or fish-lizard, is not confined to this formation, but has been found in strata as high as the chalk-marl and gault of England, and as low as the muschelkalk, a formation which immediately succeeds the lias in the descending order.*   It is evident from their fish-like vertebræ, their paddles, resembling those of a porpoise or whale, the length of their tail, and other parts of their structure, that the habits of the Ichthyosaurs were aquatic.   Their jaws and teeth show that they were carnivorous; and the half digested remains of fishes and rep- tiles, found within their skeletons, indicate the precise nature of their food.†   Mr. Conybeare was enabled, in 1824, after examining many skeletons nearly perfect, to give an ideal restoration of the osteology of this genus, and of that of the Plesio- saurus.‡   The latter animal had an extremely long neck and small head, with teeth like those of the crocodile, and paddles analogous to those of the Ichthyosaurus, but larger.   It is supposed to have lived in shallow seas and estuaries, and

* Buckland, Bridgew. Treat., p. 168.
† Ibid. p. 187.
‡ Geol. Trans., Second Series, vol. i. pl. 49.

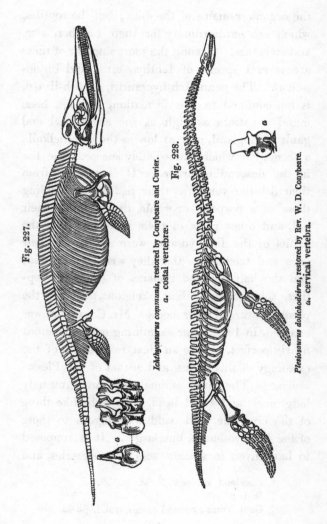

Fig. 227.

Ichthyosaurus communis, restored by Conybeare and Cuvier.
a. costal vertebra.

Fig. 228.

Plesiosaurus dolichodeirus, restored by Rev. W. D. Conybeare.
a. cervical vertebra.

to have breathed air like the Ichthyosaur, and our modern cetacea.* Some of the reptiles above mentioned were of formidable dimensions. One specimen of *Ichthyosaurus platyodon*, from the lias at Lyme, now in the British Museum, must have belonged to an animal more than twenty-four feet in length, and another of the Plesiosaurus, in the same collection, is eleven feet long. The form of the Ichthyosaurus may have fitted it to cut through the waves like the porpoise; but it is supposed that the Plesiosaurus, at least the long-necked species (Fig. 228.), was better suited to fish in shallow creeks and bays, defended from heavy breakers.

For the last twenty years, anatomists have agreed that these extinct saurians must have inhabited the sea, although no living marine reptile was known. They argued that, as there are now Chelonians, like the tortoise, living in fresh water, and others, as the turtle, frequenting the ocean, so there may have been formerly some saurians proper to salt, others to fresh water. The recent discovery, however, of a maritime saurian, has now rendered it unnecessary to speculate on such possibilities. This creature was found in the Galapagos islands, during the visit of H. M. S. Beagle to that archipelago, in 1835, and its habits

* Conybeare and De la Beche, Geol. Trans.; and Buckland, Bridgew. Treat., p. 203.

were then observed by Mr. Darwin. The islands
alluded to are situated under the equator, nearly
600 miles to the westward of the coast of South
America. They are volcanic, some of them
being 3000 or 4000 feet high; and one of them,
Albemarle Island, 75 miles long. The climate
is mild, very little rain falls; and, in the whole
archipelago, there is only one rill of fresh water
that reaches the coast. The soil is for the most
part dry and harsh, and the vegetation scanty.
The birds, reptiles, plants, and insects are,
with very few exceptions, of species found nowhere else in the world, although all partake,
in their general form, of an American character. Of the mammalia, says Mr. Darwin,
one species alone appears to be indigenous,
namely, a large and peculiar kind of mouse; but
the number of lizards, tortoises, and snakes is so
great that it may be called a land of reptiles.
The variety, indeed, of species is small; but the
individuals of each are in wonderful abundance.
There is a turtle, a large tortoise (*Testudo Indi-
cus*), four lizards, and about the same number of
snakes, but no frogs or toads. Two of the lizards
belong to the family *Iguanidæ* of Bell, and to
a peculiar genus (*Amblyrhynchus*) established by
that naturalist; and so named from their obtusely
truncated head and short snout.* Of these li-

---

* αμϐλυς, amblys, blunt, and ῥυγχος, rhynchus, snout.

zards, one is terrestrial in its habits, and burrows in the ground, swarming everywhere on the land, having a round tail, and a mouth somewhat resembling in form that of the tortoise. The other is aquatic, and has its tail flattened laterally for swimming (see Fig. 229.). " This marine saurian," says Mr.

Fig. 229.

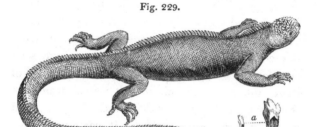

Amblyrhynchus cristatus, Bell. Length varying from 3 to 4 ft. *The only existing marine lizard now known.*
a. Tooth of same of natural size, and magnified.

Darwin, " is extremely common on all the islands throughout the archipelago. It lives exclusively on the rocky sea-beaches, and I never saw one even ten yards inshore. The usual length is about a yard, but there are some even four feet long. It is of a dirty black colour, sluggish in its movements on the land ; but, when in the water, it swims with perfect ease and quickness by a serpentine movement of its body and flattened tail, the legs during this time being motionless, and closely collapsed on its sides. Their limbs and strong claws are admirably adapted for crawling over the rugged and fissured masses of lava which everywhere form the coast. In such situations a group of six or seven of these hideous reptiles may oftentimes be seen on the black rocks, a few feet above the surf, basking in the sun with outstretched legs. Their stomachs, on being opened, were found to be largely distended with minced sea-weed, of a kind which grows at the bottom of the sea, at some little distance from the coast. To obtain this, the lizards are seen occasionally going out to sea in shoals. One of these animals was sunk in salt water, from the ship, with a heavy weight attached to it, and drawn up again after an hour; it was quite active and unharmed. It is not yet known by the inhabitants where this animal lays its eggs; a singular fact, considering its abundance, and that the natives are well acquainted with the eggs of the terrestrial *Amblyrhynchus*, which last is also herbivorous, although feeding on a very different kind of vegetation." *

In those deposits now forming by the sediment washed away from the wasting shores of the Galapagos islands, the remains of saurians, both of the land and sea, as well as of chelonians and fish, may be mingled with marine shells without any bones of land quadrupeds or batrachian reptiles;

* Darwin's Journal, chap. xix. (For full title, see note, p. 137.)

yet even here we should expect the remains of marine mammalia to be imbedded in the new strata, for there are seals, besides several kinds of cetacea, on the Galopagian shores; and, in this respect, the parallel between the modern fauna, above described, and the ancient one of the lias, would not hold good.

Sudden destruction of saurians, &c. — It has been remarked, and truly, that many of the fish and saurians, found fossil in the lias, must have met with sudden death and immediate burial; and that the destructive operation, whatever may have been its nature, was often repeated.

" Sometimes," says Dr. Buckland, " scarcely a single bone or scale has been removed from the place it occupied during life; which could not have happened had the uncovered bodies of these saurians been left, even for a few hours, exposed to putrefaction, and to the attacks of fishes and other smaller animals at the bottom of the sea." * Not only are the skeletons of the Ichthyosauri entire, but sometimes the contents of their stomachs still remain between their ribs, so that we can discover the particular species of fish on which they lived, and the form of their excrements. Not unfrequently there are layers of these coprolites at different depths in the lias, at a distance from any entire skeletons of the marine lizards from which

* Bridgew. Treat., p. 125.

they were derived, "as if," says Mr. De la Beche,
"the muddy bottom of the sea received small
sudden accessions of matter from time to time,
covering up the coprolites and other exuviæ
which had accumulated during the intervals."*
It is further stated that, at Lyme Regis, those
surfaces only of the coprolites which lay uppermost at the bottom of the sea have suffered partial
decay, from the action of water before they were
covered and protected by the muddy sediment
that has afterwards permanently enveloped them.†

Numerous specimens of the pen-and-ink fish
(*Sepia loligo*, Lin., *Loligo vulgaris*, Lam.) have
also been met with in the lias at Lyme, with the
ink-bags still distended, containing the ink in a
dried state, chiefly composed of carbon, and but
slightly impregnated with carbonate of lime.
These cephalopoda, therefore, must, like the saurians, have died suddenly, and have been in-
stantly buried in sediment; for, if exposed after
death, the membrane containing the ink would
have decayed. ‡

As we know that river fish are sometimes
stifled, even in their own element, by muddy water
during floods, it cannot be doubted that the periodical discharge of large bodies of turbid fresh
water into the sea may be still more fatal to

* Geological Researches, p. 334.
† Buckland, Bridgew. Treat., p. 307.   ‡ Ibid.

marine tribes. In the Principles of Geology, I have shown how large quantities of mud and drowned animals are swept down into the sea by rivers during earthquakes, as in Java, in 1699 ; and how undescribable multitudes of dead fish have been seen floating on the sea after a discharge of noxious vapours after similar convulsions. * But, in the intervals between such catastrophes, strata may have accumulated slowly in the sea of the lias, some being formed chiefly of one description of shell, such as ammonites, others of gryphites.

Fossil plants. — Among the vegetable remains of the Lias, several species of Zamia have been found at Lyme Regis, and the remains of coniferous plants at Whitby. Fragments of wood are common, and often converted into argillaceous limestone. That some of this wood, though now petrified, was soft when it first lay at the bottom

Fig. 230.

of the sea is shown by a specimen now in the museum of the Geological Society (see Fig. 230.), which has the form of an ammonite indented on its surface.

Origin of the Oolite and Lias. — If we now endeavour to restore, in imagination, the ancient

* See Principles, *Index*, Lancerote, Graham Island, Caabria.

condition of the European area at the period of the
Oolite and Lias, we must conceive a sea in which
the growth of coral reefs and shelly limestones,
after proceeding without interruption for ages,
was liable to be stopped suddenly by the deposition of clayey sediment. Then, again, the argillaceous matter, devoid of corals, was deposited
for ages, and attained a thickness of hundreds of
feet, until another period arrived when the same
space was again occupied by calcareous sand, or
solid rocks of shell and coral, to be again succeeded by the recurrence of another period of
argillaceous deposition.   Mr. Conybeare has remarked of the entire group of Oolite and Lias,
that it consists of repeated alternations of clay,
sandstone, and limestone, following each other in
the same order. Thus the clays of the lias are
followed by the sands of the inferior oolite, and
these again by shelly and coralline limestone,
(Bath oolite, &c.) ; so, in the middle oolite, the
Oxford clay is followed by calcareous grit and
" coral rag ;" lastly, in the upper oolite the Kimmeridge clay is followed by the Weymouth sands
and the Portland limestone. * The clay beds,
however, as Mr. De la Beche remarks, can be
followed over larger areas than the sands or sandstones. †   It should also be remembered, that

---

* Con. and Phil. p. 166.   † Geol. Researches, p. 337.

while the oolitic system becomes arenaceous, and resembles a coal-field in Yorkshire, it assumes, in the Alps, an almost purely calcareous form, the sands and clays being omitted; and even in the intervening tracts, it is more complicated and variable than appears in ordinary descriptions. Nevertheless, some of the clays and intervening limestones do, in reality, retain a pretty uniform character, for distances of from 400 to 600 miles from east to west and north to south.

According to M. Thirria, the entire oolitic group in the department of the Haute Saône, in France, may be equal in thickness to that of England; but the importance of the argillaceous divisions is in the inverse ratio to that which they exhibit in England, where they are about equal to twice the thickness of the limestones, whereas, in the part of France alluded to, they reach only about a third of that thickness.* In the Jura the clays are still thinner; and in the Alps they thin-out and almost vanish.

In order to account for such a succession of events, we may imagine, first, the bed of the ocean to be the receptacle for ages of fine argillaceous sediment, brought by oceanic currents, which may have communicated with rivers, or with part of the sea near a wasting coast. This mud ceases,

---

* Burat's D'Aubuisson, tom. ii. p. 456.

at length, to be conveyed to the same region,
either because the land which had previously suffered denudation is depressed and submerged, or
because the current is deflected in another direction by the altered shape of the bed of the ocean
and neighbouring dry land. By such changes
the water becomes once more clear and fit for the
growth of stony zoophytes. Calcareous sand is
then formed from comminuted shell and coral, or,
in some cases, arenaceous matter replaces the
clay, because it commonly happens that the finer
sediment, being first drifted farthest from coasts,
is subsequently overspread by coarse sand, after
the sea has grown shallower, or when the land,
increasing in extent, has approached nearer to the
spots first occupied by fine mud.

In order to account for another great formation, like the Oxford clay, again covering one of
coral limestone, we must suppose a sinking down
like that which is now taking place in some existing regions of coral between Australia and South
America.* The occurrence of subsidences, on so
vast a scale, may again have caused the bed of the
ocean and the adjoining land throughout the
European area, to assume a shape favourable to
the deposition of another set of clayey strata; and
this change may have been succeeded by a series

* See Darwin, chap. xxii. (for full title, see note, p. 137.)

of events analogous to that already explained, and
these again by a third series in similar order.
Both the ascending and descending movements
may have been extremely slow, like those now
going on in the Pacific; and the growth of every
stratum of coral, a few feet in thickness, may have
required centuries for its completion, during which
certain species of organic beings may have disappeared from the earth, and others have been
introduced in their place; so that, in each set of
strata, from the Upper Oolite to the Lias, some
peculiar and characteristic fossils were imbedded.

Physical geography. — The alternation, on so
large a scale, of distinct formations of clay and
limestone, has given rise to some marked features
in the physical outline of parts of England and
France. Wide valleys can usually be traced
throughout the long bands of country where the
argillaceous strata crop-out; and between these
valleys the limestones are observed, composing
ranges of hills, or more elevated grounds. These
ranges terminate abruptly on the side on which
the several clays crop-out from beneath the calcareous strata.

The annexed diagram will give the reader an
idea of the configuration of the surface now
alluded to, such as may be seen in passing from
London to Cheltenham, or in other parallel lines,
from east to west, in the southern part of Eng-

Fig 231.

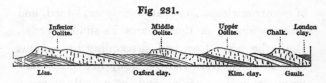

land. It has been necessary, however, in this drawing, greatly to exaggerate the inclination of the beds, and the height of the several formations, as compared to their horizontal extent. It will be remarked, that the lines of cliff, or escarpment, face towards the west in the great calcareous eminences formed by the Chalk and the Upper, Middle, and Lower Oolites; and at the base of each we have respectively the Gault, Kimmeridge clay, Oxford clay, and Lias. This last forms, generally, a broad vale at the foot of the escarpment of Inferior Oolite; but a considerable portion of that escarpment is sometimes occupied by lias. The external outline of the country which the geologist observes in travelling westward from Paris to Metz, is precisely analogous, and is caused by a similar succession of rocks intervening between the tertiary strata and the Lias; with this difference, however, that the escarpments of Chalk, Upper, Middle, and Inferior Oolites, face towards the east instead of the west.

The Chalk crops-out from beneath the tertiary sands and clays of the Paris basin, near Epernay, and the Gault from beneath the Chalk and Upper

Green-sand at Clermont en Argonne; and passing from this place by Verdun and Etain to Metz, we find two limestone ranges, with intervening vales of clay, precisely resembling those of southern and central England, until we reach the great plain of Lias at the base of the Inferior Oolite at Metz.

It is evident, therefore, that the denuding causes have acted similarly over an area several hundred miles in diameter, sweeping away the softer clays more extensively than the limestones, and undermining these last so as to cause them to form steep cliffs wherever the harder calcareous rock was based upon a more yielding and destructible clay. This denudation probably occurred while the land was slowly rising out of the sea.*

* See Princ. of Geol., *Index*, Wealden denudation.

CHAPTER XIX.

NEW RED SANDSTONE GROUP.

Distinction between New and Old Red sandstone — Between
Upper and Lower New Red — Muschelkalk in Germany
— Fossil plants and shells of New Red Group, entirely
different from Lias and Magnesian limestone — Lower
New Red and Magnesian limestone — Zechstein in Germany of the same age — General resemblance between the
organic remains of the Magnesian limestone and Carboniferous strata — Origin of red sandstone and red marl.

BETWEEN the Lias and the Coal, or Carboniferous
group, there is interposed in the midland and
western counties of England a great series of red
marls and sandstones, to which the name of the
New Red Sandstone formation was given, to distinguish it from other marls and sandstones called
the " Old Red," (*c.* Fig. 232.), often identical in
mineral character, which lie immediately beneath
the coal, *b.*

Fig. 232.

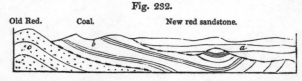

Old Red. Coal. New red sandstone.

In some parts of the south-west of England,
the entire " New Red " group consists exclusively

of red loam, clay, and sandstone, devoid of fossils, strongly contrasted in colour, and the general absence of calcareous matter, with the Oolitic rocks and Lias before described. But when we extend our observations over England and other countries, we no longer find this simplicity of structure; but perceive that the strata between the Lias and the Coal are divisible into two very distinct systems, which will be understood from the accompanying Table, and the description which follows.

NEW RED SANDSTONE GROUP.

*Poikilitic group of Conybeare and Buckland.**

		Synonyms.	
		German.	French.
1. Upper new red.	*a.* Saliferous marls and sandstone -	Keuper - -	Marnes irisées.
	b. (wanting in England) - -	Muschelkalk -	Muschelkalk, ou calcaire coquillier.
	c. Sandstone and quartzose conglomerate - -	Bunter sand-stein - -	Grès bigarré.
2. Lower new red.	*d.* Magnesian limestone (dolomitic conglomerate) -	Zechstein, and Kupfer schiefer - -	Zechstein, ou schiste cuivreux — et Calcaire Magnésien.
	e. Lower New Red sandstone.	Roth-liegen-des - -	Grès des Vosges, couches inféri-eures ?

* From ποικίλος, Poikilos, *variegated*, see Buckland, Bridgw. Treat., vol. ii. p. 38., because some of the most characteristic strata of this group were called *variegated* by Werner, from their exhibiting spots and streaks of light blue, green, and buff colour, in a red base.

UPPER NEW RED SANDSTONE

(Including the Muschelkalk of the Germans).

The Lias is succeeded in England by strata of
red and green marl, or clay, which are conformable to the Lias, and pass into it, as in Gloucestershire. It is in this upper New Red system that
rock salt and salt springs occur in Cheshire and
other parts of England; and to this, therefore,
the term " Saliferous marl and sandstone formation, is properly applicable."* It consists, in
Cheshire, of alternating beds of red and green
clay, or marl, gypsum, and rock-salt, upwards of
600 feet in thickness.

A few traces only of fossil shells, fish, and
plants have been detected in this formation in
England; but in a corresponding position in Germany there occur similar strata of red sandstone
and marl, in which are many organic remains,
and associated with the same a great calcareous formation called the " Muschelkalk," or
" shelly-limestone." As the fossil fauna and
flora of these formations supply the chasm which
exists in our British series, I shall say a few words
of the " Upper New Red," as it appears in Bavaria and Wurtemberg. First in order beneath
the Lias come mottled marls and sandstones, red,

* Murchison, Silurian System, p. 32.

green, purple, and white, containing gypsum and
salt; then the Muschelkalk above mentioned, and
then another set of marls and sandstones much
resembling the first. That these three formations,
the Keuper, Muschelkalk, and Bunter Sandstein
(see Table), may be referred to one period, appears from the fact that Count Munster has
obtained the same plants from the Keuper and
Bunter Sandstein; and M. Agassiz the same species of fish from both of them, and from the interposed Muschelkalk. It is also worthy of remark,
that the strata of the Muschelkalk alternate with
those of the Keuper and Bunter Sandstein at
their junction.

The fossil Flora, above alluded to, consists of
Cycadeæ and several genera of ferns, also extinct
coniferæ of the genus Voltzia (Ad. Brongniart)
peculiar to this period, in which even the fructification has been preserved (Fig. 233.), and a
gigantic species of Equisetum (Fig. 234.), which
is not uncommon in the Keuper sandstone.

Fig. 233.   Fig. 234.

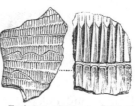

Voltzia brevifolia, and portion magnified   *Equisetum columnare;* fragment
  to show fructification; Sulzbad.   of stem, and small portion of
  Keuper and Bunter Sandstein.   same magnified. Keuper.

T

Among the shells, some of the Cephalopoda are peculiar, as, for example, that form of Ammonite which is called Ceratite by De Haan, in which the descending lobes, see *a*, *b*, *c*, Fig. 235., ter-

Fig. 235.

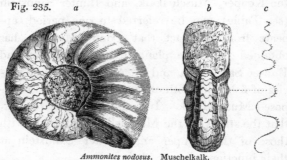

Ammonites nodosus. Muschelkalk.

a. Side view.   *b.* Front view.
c. Partially denticulated outline of the septa dividing the chambers.

minate in a few small denticulations pointing inwards. Among the bivalve shells, the *Posidonia keuperina*, Voltz. (*Posidonomya minuta*, Bronn, Fig. 236.) is abundant, ranging from the Keuper to the Bunter Sandstein; and the *Avicula socialis* (Fig. 237.), having a similar range, but most

Fig. 236.   *a*   Fig. 237.   *b*

Posidonomya   *a. Avicula socialis.*   *b. Side view of same.*
minuta, Bronn.   Characteristic of the Muschelkalk.

characteristic of the Muschelkalk in Germany, France, and Poland.

There are also some encrinites in the Muschelkalk, and some teeth of cartilaginous fish, a few decapod crustacea, and no less than five genera of large extinct reptiles, all peculiar to the Muschelkalk, as Phytosaurus, Dracosaurus, and others. Upon the whole, Professor Bronn has enumerated, in his Lethæa Geognostica, no less than forty-seven genera of fossil remains from the three divisions of the " Upper New Red" system in Germany; and these fossils are the more important as being all distinct in species, and many of them in genera, from those of the incumbent Lias or more ancient Magnesian limestone.

In the Bunter Sandstein near Hildburghausen, some remarkable fossil footsteps have lately been discovered in quarries of a grey quartzose sandstone. On the upper surfaces of the slabs of stone the steps form depressions, while those on the lower surfaces are in relief. These last are natural casts formed in the subjacent footsteps, as in moulds. The larger prints seem to be those of the hind foot, and are about eight inches long and five wide. Near to each, and at the regular distance of an inch and a half before them, is a smaller print of a fore-foot (see Fig. 238.). In each pair of large and small steps the great toes are turned alternately both to the right or both to the left.*

* One of these slabs is now in the British Museum.

T 2

Fig. 238.

Single footstep of Chirotherium,
one eighth of nat. size.

For this unknown animal Professor Kaup has proposed the provisional name of Chirotherium; and he conjectures that it was a mammiferous quadruped, allied to the marsupialia.*

In the kangaroo, says Dr. Buckland, the first toe of the

Fig. 239.

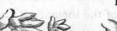

Line of footsteps on slab of sandstone.   Hildburghausen, in Saxony.

fore-foot is, in a similar manner, set obliquely to the others, like a thumb; and the disproportion between the fore and hind-feet is also very great. If it should be eventually proved that this animal was really marsupial, these fossil relics belong to the most ancient mammiferous quadruped yet known to palæontologists.

It would scarcely be possible to draw a distinct line of demarcation between the Keuper and Bunter Sandstein, in Germany, where they are not barren of fossils, if the Muschelkalk did not intervene between them. In England, therefore, where this calcareous formation is wanting, and where there are scarcely any organic remains in

* See Buckland's Bridgew., p. 263.

the Upper New Red marl and sandstone, we cannot feel assured that the divisions *a.* and *c.* of our Table, p. 407., do really coincide with the German Keuper and Bunter Sandstein. But it has been found convenient in the counties of Salop, Stafford, and Worcester, to divide the saliferous marls from the inferior quartzose conglomerate in the manner above indicated.

LOWER NEW RED SANDSTONE AND MAGNESIAN LIMESTONE.

We now come to the Lower New Red system, the position of which can best be determined in Germany, because it is there interposed between the Coal and Bunter Sandstein, or oldest part of the " Upper New Red," above described. In the south-west of England the New Red sandstone formation is unconformable to the Coal (see Fig. 232.); but in the north-east of England Professor Sedgwick has shown that the same series is conformable to the carboniferous strata, and passes into them. In other words, the movements which deranged " the Coal" in the south-west, previously to the origin of the New Red sandstone, did not extend towards Durham and the more northern counties.

Near Bristol, in Somersetshire, and in other counties bordering the Severn, the unconformable beds of the Lower New Red, resting immediately

upon the Coal, consist of a conglomerate called
" dolomitic," because the pebbles of older rocks
are cemented together by a base of magnesian
limestone.   Among the imbedded pebbles are
many derived from the Coal, particularly from
the carboniferous limestone, the peculiar fossils of
which are still seen in many large rounded fragments.   In the north-east of England the dolomitic
conglomerate is represented by a yellow limestone, generally called the Magnesian Limestone,
which passes upwards into marl slate, and downwards into red marl and gypsum.   In the intermediate counties of Worcestershire, Staffordshire,
and Shropshire, are conglomerates referred to the
same age, but which are calcareous, with scarcely
any magnesia.   Between these conglomerates and
the Coal is a great formation, called the Lower
New Red sandstone (see Table, p. 407.), composed of sandstones, red shales, and marls, occasionally spotted green.*

The country of Mansfeld, in Thuringia, may
be called the classic ground of the Lower New
Red, or Magnesian Limestone formation, on the
continent.   It has there been long celebrated, because one of its members, a slaty marlstone, is
richly impregnated with copper pyrites, for which
it is extensively worked.   The formation in that

* Murchison, Silurian System, p. 54.

country is composed of an upper calcareous division, called the Zechstein, and a lower red quartzose formation of sandstone and conglomerate, called the Rothliegendes. The upper of these systems is very complex, consisting of marl, limestone, copper-slate, magnesian limestone, gypsum, and rock-salt, in which numerous fossils occur, bearing a striking generic resemblance to those of our English Magnesian Limestone. The Lower system, or Rothliegendes, is interposed between the Zechstein and the Coal; and is supposed to correspond with the Lower New Red sandstone, above mentioned, as occupying a similar place in England between our Magnesian Limestone and Coal. Its local name of Rothliegendes, *red-lyer*, or "Roth-todt-liegendes," *red-dead-lyer*, was given by the workmen in the German mines from its red colour, and because the copper has *died out* when they reach this rock, which is not metalliferous. It is, in fact, a great deposit of red sandstone and conglomerate, with associated porphyry, basaltic trap, and amygdaloid.

When we consider the fossils of the Magnesian Limestone in England, or corresponding Zechstein in Germany, we find that they approach much nearer in their character to the organic remains of the older carboniferous group than to those of the Upper New Red. Thus, for example, the two genera of shells, Producta and Spirifer,

of the family Brachiopoda, are common to the
Magnesian Limestone, Coal, and Primary fossili-

Fig. 240.   Fig. 241.

Producta calva, Sow.   *Spirifer undulatus*, Sow.
(*Leptæna*, Dalman.)   Magnesian Limestone.
Magnesian Limestone.

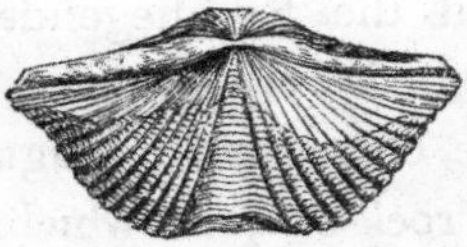

ferous strata, but have never been met with in
any rock above the Magnesian Limestone. There
are certain fish also found both in England and
Germany, in the Lower New Red system, which
occur in the carboniferous strata, but in no formation higher in the series than the Magnesian
limestone, not even in the Muschelkalk.

Fig. 242.

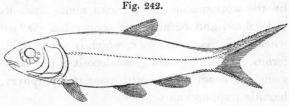

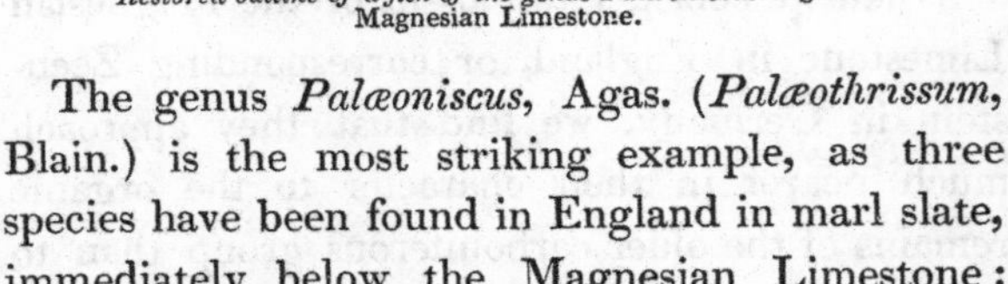

Restored outline of a fish of the genus Palæoniscus. Agass. *
Magnesian Limestone.

The genus *Palæoniscus*, Agas. (*Palæothrissum*,
Blain.) is the most striking example, as three
species have been found in England in marl slate,
immediately below the Magnesian Limestone;

* Poissons Fossiles, vol. i. tab. A. fig. 4.

and three other different, but nearly allied species, in the slate of the Zechstein of Germany.*

It was first pointed out by M. Agassiz, that all the bony fish of the Magnesian Limestone, and of all the more ancient formations, have the vertebral column continued into the upper lobe of the tail, which is much longer than the lower lobe (see Fig. 242.), whereas, in strata newer than the Magnesian Limestone, the tail-fin is divided into two equal lobes, as in almost all living fishes, the vertebræ not being prolonged into either lobe.

The remains of at least two saurian animals of new genera, Palæosaurus and Thecodontosaurus have been lately discovered in the dolomitic conglomerate near Bristol.† They are allied to the Iguana and Monitor, and are the most ancient examples of fossil reptiles yet found in Great Britain. The Zechstein of Germany is also the oldest rock on the continent in which Saurian remains have been found. They are referred to a genus called Protorosaurus, also allied to the Monitor.

The resemblance above alluded to between the fossils of the Lower New Red system and those of the Coal, is not confined to the mollusca, fish, and

* Sedgwick, Geol. Trans., Second Series, vol. iii. p. 117.
† See paper by Messrs. Riley and Stuchbury, Proceedings Geol. Soc. No. 45.

Fig. 243.

Cyathocrinites planus, Miller. Magnesian and Mountain Limestone.

reptiles, but extends to the Crinoidea, or Stone-lilies. Thus one species, the *Cyathocrinites planus* (Fig. 243.) of the Magnesian Limestone of Durham, has been identified by Mr. Miller with a fossil of the Mountain limestone of Bristol.*

Origin of the New Red Sandstone group. — The red sandstone and red marl, which, in point of thickness, form the most considerable part both of the upper and lower New Red formation in England and Germany, may have arisen in great part from the disintegration of various crystalline, or metamorphic schists; and sometimes, as in parts of Saxony and Devonshire, from porphyritic trap rocks containing much oxide of iron. In some districts of the eastern Grampians in Scotland, as in the north of Forfarshire, the sides of mountains composed of gneiss, mica-schist, and clay-slate, are covered with alluvium, derived from the disintegration of those rocks; and the mass of detritus is stained by oxide of iron, of precisely the same colour as the Old Red sandstone of the adjoining Lowlands. Now this alluvium merely requires to be swept down to the sea, or into a

* Sedgwick, Geol. Trans., Second Series, vol. iii. p. 120.

lake, to form strata of red sandstone and red marl, similar to those of the " Old Red" or New Red system, or those of the cretaceous era in Spain (see p. 343.), or those of tertiary origin, as at Coudes and Champheix, in Auvergne, all of which are in lithological characters quite undistinguishable from one another. The pebbles of gneiss in the tertiary red sandstone of Auvergne, point clearly to the rocks from which it has been derived. The red colouring matter may have been furnished by the decomposition of hornblende, or mica, which contain oxide of iron in large quantity (see p. 167.).

It is a general fact, and one not yet accounted for, that scarcely any fossil remains are preserved in stratified rocks in which this oxide of iron abounds; and when we find fossils in the New or Old Red sandstone in England, it is in the grey, and usually calcareous beds, that they occur.

CHAPTER XX.

THE COAL, OR CARBONIFEROUS GROUP.

Carboniferous strata in the south-west of England — Superposition of Coal-measures to Mountain limestone — Departure from this type in north of England and Scotland —
Freshwater strata — Intermixture of freshwater and marine
beds — Sauroidal fish — Fossil plants — Ferns and Sigillariæ — Lepidodendra — Calamites — Coniferæ — Stigmariæ.

THE next group which we meet with in the descending order is the Carboniferous, commonly
called " The Coal," because many beds of that
mineral, in a more or less pure state, are interstratified with sandstone, shale, and limestone, of
which the bulk of the formation is made up. The
combustible coal itself, even in Great Britain and
Belgium, where it is most abundant, constitutes
but a small proportion of the whole mass. In the
north of England, for example, the thickness of
the coal-bearing strata has been estimated at 3000
feet, while the various coal-seams, 20 or 30 in
number, do not exceed 60 feet.*

In the south-west of England, in Somerset-

* Phillips; art. " Geology," Encyc. Britan.

shire, and in South Wales, the Carboniferous series consists of,

1st. Coal-measures.	Strata of shale, sandstone, and grit, with occasional seams of coal, sometimes exceeding 600 feet in thickness.
2d. Millstone grit.	A coarse quartzose sandstone passing into a conglomerate, sometimes used for mill-stones; devoid of coal; occasionally above 600 feet thick.
3d. Mountain or Carboniferous limestone.	A calcareous rock containing marine shells and corals, devoid of coal; thickness variable; sometimes 900 feet.

Beneath all these is the Old Red sandstone, which was formerly considered as part of the Carboniferous series; but which, now that its organic remains are better known, appears entitled to rank as a distinct formation.

As we proceed northwards from South Wales and Somersetshire to Yorkshire and the more northern counties, we find the Carboniferous group beginning gradually to assume a new character, there being first a slight intermixture of the Coal-measures and Mountain limestone at their contact, and these alternations taking place afterwards on a still greater scale. The Coal, in Yorkshire, does not cease when we reach the Millstone-grit, although it is there in diminished quantity; and beneath that grit is a complex deposit, 1000 feet thick, of limestones, alternating with coal-bearing sandstones and shale, below which comes the great mass of mountain limestone.* In Scotland we observe a still wider departure from the type of the south of England, the mixture of marine limestone with sandstone and shale, containing coal being more complete.

The importance of the coal in England, considered economically, is greatly enhanced by the rich beds of iron-ore which occur in the associated shales, and the contiguity of the mountain limestone which is required as a flux to reduce the iron-ore to a metallic state. †

It is now generally admitted, that all coal is of vegetable origin, the vegetable structure being still recognizable in many kinds of coal, when slices thin enough to transmit light are obtained and examined by the microscope. Impressions also of plants, together with entire trunks of trees, are frequently met with in the accompanying shale and sandstone; leaves also, and small branches, and fruits, occur in nodules of clay-ironstone, the inclosed vegetable having served as a nucleus round which the ferruginous matter, usually carbonate of iron, has concreted. Some of the coal-measures are of freshwater origin, and may have been formed in lakes, others seem to have been de-

* Sedgwick, Geol. Trans., Second Series, vol. iv. ; and Phillips, Geol. of Yorksh., part 2.

† Conybeare, Outlines, &c., p. 333.

posited in estuaries, or at the mouths of rivers, in spaces alternately occupied by fresh and salt water.

Thus a freshwater deposit, near Shrewsbury, has been ascertained by Mr. Murchison to be the youngest member of the carboniferous series of that district, at the point where the coal-measures pass into the lower New Red formation. It consists of shales and sandstones about 150 feet thick, with coal, and traces of plants, including a bed of limestone, varying from two to nine feet in thickness, which is cellular, and resembles the lacustrine limestone of France and Germany. It has been traced for 30 miles in a straight line, and recognized at more distant points. The characteristic fossils are a small bivalve, having the form of a cyclas, a small cypris, (Fig. 245.) and a microscopic shell, (microconchus) of an extinct genus.

Freshwater fossils. — Coal.
Fig. 244.   Fig. 245.

a. *Microconchus carbonarius,*
b. var. of same nat. size, and magnified.

Cypris inflata, natural size, and magnified. Murchison.*

* Silurian System, p. 84.

But in the lower coal-measures of Coalbrook Dale, the strata, according to Mr. Prestwich, often change completely within very short distances, beds of sandstone passing horizontally into clay, and clay into sandstone. The coal-seams often wedge out or disappear; and sections, at places nearly contiguous, present marked lithological distinctions. In this single field, in which the strata are from 700 to 800 feet thick, between 40 and 50 species of terrestrial plants have been discovered, besides several fishes and trilobites; the latter distinct in form from those occurring in the Silurian strata. Also upwards of 40 species of mollusca, among which are two or three of the freshwater genus Unio, and others of marine forms such as Nautilus, Orthoceras, Spirifer, and Productus. Mr. Prestwich suggests, that the inter-mixture of beds containing freshwater shells with others full of marine remains, and the alternation of coarse sandstone and conglomerate with beds of fine clay or shale containing the remains of plants, may be explained by supposing that the deposit of Coalbrook Dale, originated in a bay of the sea or estuary into which flowed a considerable river subject to occasional freshes.*

In the Edinburgh coal-field at Burdiehouse, fossil fishes, mollusca and cypris, very similar to

* Prestwich, Geol. Soc. Proceedings, No. 46. Murchison, Silurian System, p. 105.

those in Shropshire and Staffordshire, have been found by Dr. Hibbert.* In the coal-field also of Yorkshire there are freshwater strata, some of which contain shells referred to the genus Unio; but in the midst of the series there is one thin but very widely spread stratum, abounding in marine shells, such as *Ammonites Listeri* (Fig. 246.) *Orthoceras, Pecten papyraceus* (Fig. 247.), and several fishes. †

Fig. 246.

Ammonites Listeri, Sow.

Fig. 247.

Pecten papyraceus, Sow.

No similarly intercalated layer of marine shells has been noticed in the neighbouring coal-field of Newcastle, where, as in South Wales, and Somersetshire, the marine deposits are entirely below those containing terrestrial and fresh-water remains.‡

No bones of mammalia or reptiles have as yet been discovered in strata of the carboniferous group. The fish are numerous, and for the most part very remote in their organization from those

* Trans. Roy. Soc. Edin. vol. xiii.   Horner, Edin. New Phil. Journ., April, 1836.

† Phillips; art. "Geology," Encyc. Metrop., p. 590.

‡ Ibid., p. 592.

now living, as they belong chiefly to the Sauroid
family of Agassiz; as Megalichthys, Holopty-
chus, and others, which were often of great size,
and all predaceous. Their osteology, says M.
Agassiz, reminds us in many respects of the skele-

Fig. 248.

Megalichthys Hibberti, Ag.
Edinburgh coal-field;
natural size.

tons of saurian reptiles, both
by the close sutures of the
bones of the skull, their large
conical teeth striated longitudinally (see Fig. 248.), the articulations of the spinous processes with the vertebræ, and
other characters. Yet they do
not form a family intermediate
between fish and reptiles, but
are true *fish.* *

The annexed figure repre-
sents a large tooth of the Megalichthys, found by
Mr. Horner in the Cannel coal of Fifeshire. It
probably inhabited an estuary, frequenting both
the mouths of rivers and the sea.

Fossil Plants of the Coal. — But the flora of the
coal forms the most interesting feature in its palæontology, and is far better known to us than
any other flora antecedent to the tertiary era.
About 300 species of terrestrial plants are enumerated by M. Adolphe Brongniart as proper to

* Agassiz, Poiss. Foss., livr. 4. p. 62. and livr. 5. p. 88.

the Coal, but botanists have encountered the greatest difficulty in determining the natural affinities of these fossils, it being rare to find in them any vestige of flower, seed, or fruit, those organs which afford the most convenient characters for classifying living plants. They have been obliged, therefore, first to study more minutely the different forms of bark in existing trees, their various modes of branching, the tissue of their wood, nervures of the leaves, and other peculiarities of vegetable structure which might enable them to institute a direct comparison between the analogous parts of recent and fossil plants.*

The most common of these vegetable remains may be provisionally classed under the following heads : — First, Ferns and Sigillariæ; secondly, Lepidodendra, allied to *Lycopodiaceæ?* thirdly, Calamites, allied to *Equisetaceæ?* fourthly, Coniferous plants; fifthly, Stigmariæ, apparently an extinct family of plants.

Ferns and Sigillariæ. — The leaves, or more properly speaking the fronds, of ferns (see Figs. 249, 250.), for the most part destitute of fructification, exceed in number all other plants in the shale of the coal. They have been divided by M. Ad. Brongniart into genera, characterized chiefly by the branching of the fronds, and the way in which

* See the works of MM. Ad. Brongniart, Sternberg, and others, and the Fossil Flora of Lindley and Hutton.

the veins of the leaves are disposed. These fronds
are often accompanied by large fluted stems or

Fig. 249.   Fig. 250.

Pecopteris lonchitica.
(Foss. Flo. 153.)

a. Sphenopteris crenata.
b. The same, magnified.
(Foss. Flo. 101.)

trunks of trees which have been squeezed down
and flattened as they lay prostrate in the shale,
so that the opposite sides meet, but which when
they occur in the accompanying grit or sandstone,
and are placed obliquely or vertically to the planes
of stratification, are round and uncompressed.
Their bark has been converted into coal; and
they must have been hollow when first deposited,
for the interior became filled, not only with sand,
but with leaves and branches of ferns, introduced
from above. Impressions of these fronds are

now frequent in the pillars of sandstone, which may be regarded as casts of the interior of those ancient trees. Most of the trunks or stems now alluded to have been called Sigillariæ. They vary from half a foot to five feet in diameter, and must have been sometimes forty or fifty feet high.

It is admitted by all botanists that some of these gigantic stems, all of which are comprehended by Brongniart in his genus Sigillaria, were true arborescent ferns, as for example, that section which has been named Caulopteris by Lindley and Hutton. (see Fig. 251.) But these are comparatively rare, whereas of the other section (Fig.

Fig. 251.   Fig. 252.

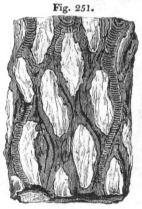
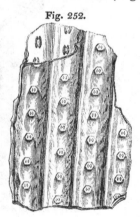

Sigillaria Lindleyi, Brong.   *Sigillaria lævigata*, Brong.
(*Caulopteris primæva*, Lindley.)

252.) more than forty species have been described. In these the scars on the stem are smaller and more regularly arranged in parallel series on the fluted bark (Fig. 252.)

The recent tree-ferns belong to one tribe (*Polypodiaceæ*), a..d to a small number only of genera in that tribe, in all of which the surface of the trunk is marked with scars, or cicatrices, left after the fall of the fronds.   These scars are precisely similar to those of Caulopteris (Fig. 251.) ; but Mr. Lindley objects to the opinion that the remaining Sigillariæ of Brongniart were Tree-ferns,

Fig. 253.   Fig. 254.   Fig. 255.

Living Tree Ferns of different genera. (Ad. Brong.)

Fig. 253. Tree fern from Isle of Bourbon.
Fig. 254. *Cyathea glauca*, Mauritius.
Fig. 255. Tree fern from Brazil.

because the scars in these are smaller, dissimilar in form, and more regularly arranged in parallel

lines; also, because the stems are fluted (see
Fig. 252.), and sometimes bifurcating. M. Brongniart has replied, that the forking of the stems of
some of the fossil trees is no more than might
have been expected from their large size; and as
to the forms of the discs or scars from which the
fronds have fallen, their individual variations are
not greater than those which we find in the fronds
of different genera of living ferns, which do not
in the present state of the globe attain the size of
trees.

Lepidodendra. — Another class of fossils, very
common in the coal-shales, have been named Lepidodendra. Some of these are of small size, and

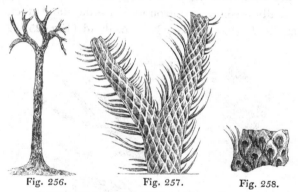

Fig. 256.   Fig. 257.   Fig. 258.

Lepidodendron Sternbergii. Coal-measures, near Newcastle.

Fig. 256. Branching trunk, 49 feet long, supposed to have belonged to *L. Sternbergii.* (Foss. Flo. 203.)

Fig. 257. Branching stem with bark and leaves of *L. Sternbergii.*
(Foss. Flo. 4.)

Fig. 258. Portion of same nearer the root; natural size. (Ibid.)

approach very near in form to the modern *Lyco-podiums*, or club-mosses, while others of much larger dimensions are supposed to have been intermediate between these and coniferous plants. The annexed figures represent a large fossil, Lepidodendron, forty-nine feet long, lately found in Jarrow Colliery, near Newcastle, lying in shale parallel to the planes of stratification. Fragments of others, found in the same shale, indicate by the size of the rhomboidal scars which cover them a still greater magnitude. The living club-mosses, of which there are about 200 species, are abundant in tropical climates, where one species is sometimes met with attaining a height of three feet. They usually creep on the ground, but some stand erect, as the *L. densum*, from New Zealand (Fig. 259.).

Fig. 259.

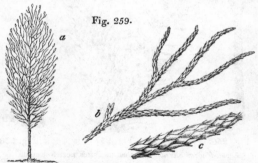

a. Lycopodium densum ; banks of R. Thames, New Zealand.
b. branch, natural size.   *c.* part of same, magnified.

Calamites. — These fossils have a jointed stem longitudinally striated, and are supposed by M. Brongniart to have been allied to the *Equisetaceæ*,

or horse-tail tribe; aquatic plants which, in a living
state, are only two or three feet high in our climates, and even in tropical countries only attain,
as in the case of *Equisetum giganteum*, discovered
by Humboldt and Bonpland, in South America, a
height of about five feet, the stem being an inch
in diameter. The Calamites, however, of the Coal
differed from these, principally in being furnished
with a thin bark, which is represented in the stem
of *C. Suckowii* (Fig. 261.), in which it will be seen

Fig. 260.   Fig. 261.

Calamites cannæformis, Schlot.
(Foss. Flo. 79.) Common in
English coal.

Calamites Suckowii, Brong.
natural size. Common in
coal throughout Europe.

that the striped external pattern does not agree
with that left on the stone where the bark is
stripped off, so that if the two impressions were
seen separately, they might be mistaken for two
distinct species.

Coniferæ. — The structure of the wood of certain coal plants displays so great an analogy to
that of certain pines of the genus Araucaria, as to

lead to the opinion that some species of firs existed at this period. (See above, p. 82.)

Stigmariæ. — Fragments of a plant which has been called *Stigmaria ficoides* occur in great numbers in almost every coal-pit. It is supposed to have been a huge succulent water-plant of an extinct family; thin transparent sections of the stem exhibiting an anatomical structure quite different from the wood of any living tree.*   According to

Fig. 262.

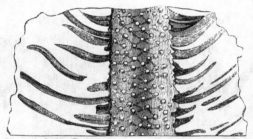

Stigmaria ficoides, Brong.   One fourth of nat. size.   (Foss. Flo. 32.)

Fig. 263.

Surface of another individual of same species, showing form of tubercles. (Foss. Flo. 34.)

the conjectures of some botanists, it approached most nearly to the family *Lycopodiaceæ;* according to others to *Euphorbiaceæ.* Mr. Hutton discovered one of these Stigmariæ forming a huge dome-shaped body, from which twelve branches spread horizontally in all directions, each, usually dividing into two arms, from twenty

* Lindley, Foss. Flora, p. 166.

to thirty feet long, to which leaves of great length were attached.   Dr. Buckland imagines these plants to have grown in swamps, or to have floated in lakes like the modern Stratiotes.*

I shall postpone some general remarks on the climate of the Carboniferous period, arising out of the contemplation of its flora, until something has been said of the contemporaneous Mountain lime-stone and its marine fossils.

* Bridgew. Treat., p. 478.

CHAPTER XXI.

CARBONIFEROUS or Mountain limestone. — We
have already seen that this rock lies sometimes
entirely beneath the Coal-measures, while, in other
districts, it alternates with the shales and sandstone
of the Coal. In both cases it is destitute of land
plants, and usually charged with corals, which are
often of large size; and several species belong to
the lamelliferous class of Lamarck, which enter
largely into the structure of coral reefs now growing. There are also a great number of Crinoidea
and a few Echinida, associated with the zoophytes
above mentioned. The *Brachiopoda* constitute a
large proportion of the Mollusca, many species
being referable to two extinct genera, Spirifer (or

Spirifera) (Fig. 264.) and Producta (Fig. 265.).
There are also many univalve and bivalve shells

Fig. 264.   Fig. 265.

Spirifera glabra, Sow.*
Mountain limestone.

Producta Martini, Sow.†
Mountain limestone.

of existing genera in the Mountain limestone,
such as Turritella, Buccinum, Patella, Isocardia,
Nucula, and Pecten.‡   But the Cephalopoda depart, in general, more widely from living forms,
some being generically distinct from all those
found in strata newer than the Coal.   In this
number may be mentioned Orthoceras, a siphun-
cled and chambered shell, like a Nautilus uncoiled
and straightened.   Some species of this genus are
several feet long (Figs. 266, 267.).   The Gonia-

Fig. 266.   Fig. 267.

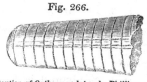

Portion of Orthoceras laterale, Phillips.
Mountain limestone.

O. giganteum, Sow.
Section showing the siphuncle;
reduced two thirds.

tite is another genus, nearly allied to the Ammonite, from which it differs in having the lobes of

* Phillips, Geol. of Yorksh. pl. 10. fig. 11.
† Ibid., pl. 8. fig. 19.   ‡ Ibid., vol. ii. p. 208.

the septa free from lateral denticulations, or cre-
natures; so that the outline of these is continuous
and uninterrupted (see *a*, Fig. 268.).   Their
siphon is small, and in the form of the striæ of
growth they resemble Nautili.   Another extinct
generic form of Cephalopod, abounding in the
Mountain limestone, and not found in strata of
later date, is the Bellerophon (Fig. 269.), of

Fig. 268.   *a*   Fig. 269.

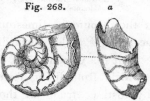

Goniatites evolutus, Phillips. *   *Bellerophon costatus*. Sow. †
Mountain limestone.   Mountain limestone.

which the shell, like the living Argonaut, was
without chambers.

Climate of the Carboniferous period. — The abundance of lamelliferous and other corals, of large
chambered Cephalopods and Crinoidea, naturally
lead us to infer that the waters of the sea, at this
period, were of a far warmer and more equable
temperature than is now experienced in those
latitudes where the Coal strata abound, in Europe.
M. Adolphe Brongniart has been led to a similar
conclusion in regard to the temperature of the
air, from considering the Carboniferous flora.

* Phillips, Geol. of Yorksh., pl. 20. fig. 65.
† Ibid., pl. 17. fig. 15.

The unquestioned existence of large tree-ferns, such as Caulopteris (Fig. 251.), now exclusively the inhabitants of hot and humid climates, and the great variety of fossil fronds of ferns in the Coal confirm this idea, even if we refuse to accede to the arguments adduced to prove that Sigillariæ were tree-ferns of extinct genera. The same views receive farther countenance, if the Lepidodendra and Calamites are rightly conjectured to have been gigantic plants of the orders *Lycopodiaceæ* and *Equisetaceæ*, which, although most largely developed at present in the tropical zone, are even there of pigmy stature in comparison with the fossil tribes just alluded to. The Araucaria, also, is a family of pines now met with in temperate and warm latitudes; and the fir trees proper to the forests of arctic regions do not appear to have any fossil representatives in the Coal. M. Ad. Brongniart, when endeavouring to establish the great heat and moisture of the climate of the era under consideration, may perhaps have relied too much on the numerical preponderance of ferns over other orders of coal-plants. We may easily be deceived by such reasoning, because it is founded on negative facts, or the absence of plants of certain orders, families, and genera. On this subject Professor Lindley has observed, that the small variety in the forms of each fossil flora must, in a great degree, depend on the relative

destructibility of plants when suspended in water
before they are imbedded in strata. In illustration of this point, he threw into a vessel containing fresh water 177 plants, among which were species of all the orders found in the Carboniferous
flora, with others representing the remaining
families and natural orders in the living creation,
and found that, at the end of two years, all had
decayed and disappeared except the ferns, palms,
Lycopodiacæ, and *Coniferæ*. The fructification of
the ferns had also vanished, but the form and
nervures of the leaves remained.*

No inference, however, drawn from this experiment, can entirely explain away the fact of the
vast preponderance in the coal-shales of fern-
leaves over those of Dicotyledonous plants. Impressions of these last, together with their wood,
are plentifully preserved in tertiary rocks in
which fossil ferns are rare; and had they been
drifted down in as large numbers as ferns into
the estuaries of the Carboniferous period, they
would have left impressions of their shape in
shale and sandstone, as they have done in more
recent formations.

It would, moreover, be rash to assume that the
coal-plants in general floated about in water for a
year or two before they were enveloped in sedi-

* Lindley, Foss. Flora, part 17.

ment. It is more probable that a large part of them were deposited immediately with the mud and sand swept down with them by rivers into lakes or the sea. This must have happened in those rare cases where the ferns still retain their fructification. Where this has disappeared, its decomposition may often have been subsequent to the inclosure of the frond in mud or sand.

Origin of the Coal strata. — Detached portions of the ancient Carboniferous group extend from Central Europe to Melville Island and the confines of the arctic region; but do not appear in the south of Europe; for the lignite and coal found south of the Alps and Pyrenees, in Spain, Italy, Greece, and other countries bordering the Mediterranean, seem referable to the Cretaceous and other comparatively modern groups.

It has been already shown that, in some parts of England, as in Shropshire, certain Coal-measures consist of freshwater strata, and may have originated in a lake, while others, not far distant, were deposited in estuaries to which the sea obtained access occasionally; while a third class were formed at the bottom of an open sea, or in bays of salt water into which land plants were drifted.*

In many parts of France and Germany there are isolated patches of Coal strata, entirely free

* Murchison, Silurian System, p. 148.

from marine fossils, which repose on granite and other hypogene rocks. They are often confined to an extremely small area, as at St. Etienne, in the department of the Loire; at Brassac, in that of Puy de Dome; at Sarrebruck; also in Silesia; and a hundred other places. All these deposits may have been formed in lakes, existing in the islands of that sea in which the Mountain limestone was formed.*

If the climate of New Zealand and the surrounding ocean was warmer, so that tree-ferns could thrive more luxuriantly on the land, and corals build reefs in the sea, we might conceive new strata to accumulate in that part of the globe analogous to those of the ancient Coal. The two islands of New Zealand are between 800 and 900 miles in length; and through the middle of them runs a lofty chain of mountains, said, in some parts, to be 14,000 feet high, and covered with perpetual snow. Many rivers descend from their sides; and, in the spring, these are copiously charged with sediment, and with abundance of drift wood. Opposite the mouths of these rivers, and near the shores, wherever these may be wasting by the action of the waves, an irregular zone of gravel, sand, and mud, must be forming in the surrounding sea — a zone several thousand

* Burat's D'Aubuisson, tom. ii. p. 268.

miles in circumference.   No less than 57 species
of ferns, some few of them arborescent, have been
already discovered in this country; and what is
remarkable, one tree-fern ranges in this country
as far south as the 46th degree, south latitude.
There are no indigenous mammalia except one
rat, and a species of bat; few reptiles, and none of
large size; so that we may anticipate a total
absence of the bones of land quadrupeds, and a
scarcity of those of reptiles, in the modern estuary
and lacustrine deposits of this region.   That there
are lacustrine strata now in progress is certain,
since one lake, called Rotorua, in the interior of
the northern island, is said to be 40 miles long,
and receives the waters of many small rivers and
torrents. *

The minor repetitions of alternate fresh and
saltwater strata in the Coal, have been ascribed to
such changes as may annually occur near the
mouths of rivers; but when shale and grit, containing coal and freshwater shells, are covered by
large masses of coralline rock, and these again by
other Coal-measures, we must suppose great
movements of elevation and subsidence, like those
by which I endeavoured to explain, in Chapters
XVI. and XVIII., the superposition of the Cretaceous group to the Wealden, or the alterna-

* Account of New Zealand, published for New Zealand
Association.

tions of argillaceous and calcareous rocks in the Oolite. In adopting such views, we must suppose the lapse of vast periods of time; as the thickness of the Coal strata, in some parts of England, independently of the Mountain limestone, has been estimated at 3000 feet. Besides, we can by no means presume that all coal-fields were in progress at once, much less that, in the same field, each mass of strata which is parallel, or occupies a corresponding level, was formed simultaneously. It is far more consistent with analogy to suppose that rivers filled up first one part of a fiord, gulf, or bay, nearest the land, and then another; so that the sea was gradually excluded from certain spaces which it previously occupied. This is doubtless the cause why the coal-bearing strata are generally uppermost, and the Mountain limestone the lowest part of each series; and why, in certain districts in the S. W. of England, the Mountain limestone suddenly thins out, so that coal-shales and grit rest immediately upon older and unconformable rocks.

Erect position of fossil trees in the Coal strata. — A great number of the fossil trees of the Coal are in a position either oblique or perpendicular to the planes of stratification. This singular fact is observed on the Continent as well as in England, and merits great attention, not only as opening a curious field for speculation, but because it has furnished a popular argument to some writers who desire to prove the earth's crust to be no more than 5000 or 6000 years old. The fact did not escape the notice of Werner, who conceived that the trees must have lived on the spots where they are now found fossil ; and this hypothesis was defended by M. Alexandre Brongniart, in the account given by him, in 1821, of the coal-mine of Treuil, at St. Etienne, near Lyons.* (Fig. 270.)

Fig. 270.

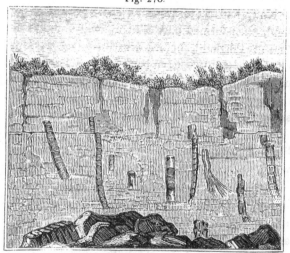

Section showing the erect position of fossil trees in coal-sandstone at St. Etienne. (Alex. Brongniart.)

In this mine, horizontal Coal strata are traversed by vertical trunks of Monocotyledonous vegetables

* Annales des Mines, 1821.

resembling bamboos, or large Equiseta. These beds are represented in the above figure (270.), and are from 10 to 13 feet in height, consisting of micaceous sandstone, distinctly stratified, and passing into the slaty structure. Since the consolidation of the stone, there has been here and there a sliding movement, which has broken the continuity of the stems, throwing the upper parts of them on one side, so that they are often not continuous with the lower.

Now, had these trees, as some geologists contend, once formed part of a submerged forest like that of Portland, before described (see p. 353.), all the roots would have been in the same stratum, or would have been confined to certain levels, and not scattered irregularly through the mass. Besides, when the stems have any roots attached to them, which happens but rarely, they are imbedded in sandstone precisely similar to that in which the trunks are inclosed, there being no soil of different composition like the Portland dirt-bed, — no line of demarcation, however slight, between the supposed ancient surface of dry land and the sediment now enveloping the trees.

Some may, perhaps, think it superfluous to advance such objections to M. Brongniart's theory, since Dr. Buckland has informed us that, when he visited these same quarries of Treuil in 1826, he saw so many trunks in an inclined posture, that

the occasional verticality of others might be accidental.* Nevertheless, the possibility of so many
of them having remained in an upright posture
demands explanation; and there are analogous
cases on record respecting similar fossils in Great
Britain of a still more extraordinary nature.

In a colliery near Newcastle, say the authors of
the Fossil Flora, a great number of Sigillarias
were placed in the rock as if they had retained
the position in which they grew. Not less than
30, some of them 4 or 5 feet in diameter, were
visible, within an area of 50 yards square, the
interior being sandstone, and the bark having
been converted into coal. The roots of one individual were found imbedded in shale; and the
trunk, after maintaining a perpendicular course
and circular form, for the height of about 10 feet,
was then bent over so as to become horizontal.
Here it was distended laterally, and flattened so
as to be only one inch thick, the flutings being
comparatively distinct.† Such vertical stems are
familiar to our miners, under the name of coal-
pipes. One of them, 72 feet in length, was discovered, in 1829, near Gosforth, about five miles
from Newcastle, in coal-grit, the strata of which
it penetrated. The exterior of the trunk was
marked at intervals with knots, indicating the

* Bridgew. Treat., p. 471.
† Lindley and Hutton, Foss. Flo., part 6. p. 150.

points at which branches had shot off. The wood
of the interior had been converted into carbonate
of lime; and its structure was beautifully shown
by cutting transverse slices, so thin as to be transparent. (See p. 82.)

In 1830, a slanting trunk was exposed in
Craigleith quarry, near Edinburgh, the total
length of which exceeded 60 feet. Its diameter
at the top was about 7 inches, and near the base
it measured 5 feet in its greater, and 2 feet in its
lesser width. The bark was converted into a
thin coating of the purest and finest coal, forming
a striking contrast in colour with the white quartzose sandstone in which it lay. The annexed

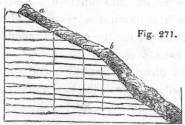

Fig. 271.

Inclined position of a fossil tree cutting through horizontal beds of sandstone,
Craigleith quarry, Edinburgh. Angle of inclination from a to b 27°.

figure represents a portion of this tree, about 15
feet long, which I saw exposed in 1830, when all
the strata had been removed from one side. The
beds which remained were so unaltered and undisturbed at the point of junction, as clearly to
show that they had been tranquilly deposited

round the tree, and that the tree had not subsequently pierced through them, while they were
yet in a soft state.   They were composed chiefly
of siliceous sandstone, for the most part white; and
divided into laminæ so thin, that from six to fourteen of them might be reckoned in the thickness
of an inch.   Some of these thin layers were dark,
and contained coaly matter; but the lowest of the
intersected beds were calcareous.   The tree could
not have been hollow when imbedded, for the interior still preserved the woody texture in a perfect state, the petrifying matter being, for the
most part, calcareous. *   It is also clear, that the
lapidifying matter was not introduced laterally
from the strata through which the fossil passes, as
most of these were not calcareous.   It is well
known that, in the Mississippi and other great
American rivers, where thousands of trees float
annually down the stream, some sink with their
roots downwards, and become fixed in the mud.
Thus placed, they have been compared to a lance
in rest; and so often do they pierce through the
bows of vessels which run against them, that they
render the navigation extremely dangerous.   But
the vertical coal-plants did not always retain their
roots.   Perhaps they sank with their larger end
downwards, because the specific gravity of the

* See figures of texture, Witham, Foss. Veget., pl. 3.

wood may have been greatest near the lower end. In trees of the Endogenous class, in particular, the wood of the inferior and older part of the trunk is more dense than the upper and younger portions; and if the former should become waterlogged while the upper part of the stem still remained nearly as light as water, or even lighter, not only would the whole trunk descend perpendicularly, but when it reached the bottom it might stand upright, provided a very slight support was afforded to its lower extremity by penetrating to the depth of a foot or two into soft mud. How long such trunks, if constantly submerged, might resist decomposition, is a question which cannot, perhaps, be determined; but, judging from the duration of wooden piles constantly covered by water, and trees naturally submerged, like those in Louisiana*, we may conclude that they might endure for many years, so that their envelopment in strata, like those of the Coal, may have been effected without a very rapid rate of deposition.

If, however, we assume that strata 30 or 40 feet thick were often thrown down in a few years, months, or even days, this fact affords no ground for calculating the time required for the formation of a wide coal-field.

* See Principles, *Index*, " Bistineau."

Suppose, for example, the structure of a coal-field always resembled that exhibited in the annexed section (Fig. 272.), we might then infer, that if the lowest set of strata, *a*, having a thickness of fifty feet, required half a century for its accumulation, the strata, *a*, *b*, *c*, constituting the entire coal-field, and being 150 feet thick,

Fig. 272.

might have been completed in a century and a half. But as the beds are wedge-shaped, and often thin out ; and as the successive beds of a single coal-field are usually arranged in the form of *a*, *b*, *c*, *d*, *e* (Fig. 273.), we cannot calculate

Fig. 273.

their number from considering any one section. The deposits, *a*, *b*, *c*, *d*, *e*, traced in a given direction, may have taken each fifty years for their deposition ; but they may have been as limited in breadth as in length. They may have constituted originally a narrow strip of land like part of the delta formed by the Mississippi, since New Orleans was built, by the incessant discharge of mud and drift timber into the Gulf of Mexico. Although by this means a narrow tongue of land has been made to protrude for several leagues into the sea, yet thousands of years may elapse before a square area of low land, having a diameter of as many leagues, can be gained from the Gulf of Mexico.

OLD RED SANDSTONE.

It was stated that the Carboniferous formation was surmounted by one called the " New Red Sandstone," and underlaid by another called the Old Red, which last was formerly merged in the Carboniferous system, but is now found to be distinguishable by its fossils. The Old Red Sandstone is of enormous thickness in Herefordshire, Worcestershire, Shropshire, and South Wales, where it is seen to crop out from beneath the Coal-measures and to repose upon the Silurian rocks. In that region its thickness has been estimated by Mr. Murchison at no less than 10,000 feet. It consists there of

1st. A quartzose conglomerate passing downwards into chocolate-red and green sandstone and marl.

2d. Cornstone and marl (red and green argillaceous spotted marls, with irregular courses of impure concretionary limestone, provincially called Cornstone, mottled, red, and green; remains of fishes).

3d. Tilestone (finely laminated hard reddish or green micaceous or quartzose sandstones, which split into tiles; remains of mollusca and fishes).

I have already observed that fossils are rare in marls and sandstones, in which the red oxide of iron prevails; in the Cornstone, however, of the counties above-mentioned, fishes of the genera Ce-

phalaspis and Onchus have been discovered.*   In
the Tilestones also, Icthyodorulites, of the genus
Onchus, have been obtained; and a species of Dip-
terus, with mollusca of the genera Avicula, Arca,
Cucullæa, Terebratula, Lingula, Turbo, Trochus,
Turritella, Bellerophon, Orthoceras, and others.†

By consulting geological maps, the reader will
perceive that from Wales to the north of Scotland,
the Old Red sandstone appears in patches, and
often in large tracts.   Many fishes have been
found in it at Caithness ‡, and various organic
remains in the northern part of Fifeshire, where
it crops out from beneath the Coal formation, and
spreads into the adjoining southern half of Forfarshire ; forming, together with trap, the Sidlaw
hills and valley of Strathmore. (See section, p. 99.)
A large belt of this formation skirts the southern borders of the Grampians, from the sea-coast
at Stonehaven and the Frith of Tay to the opposite
western coast of the Frith of Clyde. In Forfarshire,
where, as in Herefordshire, it is many thousand feet
thick, it may be divided into three principal masses : 1st, red and mottled marls, cornstone and sandstone ; 2d, Conglomerate, often of vast thickness;
3d, Tilestones and paving stone, highly micaceous,
and containing a slight admixture of carbonate of

---

* Murchison's Silurian System, p. 180.   † Ibid., p. 183.
‡ See Geol. Trans. 2d series, vol. iii. plates 15, 16, 17.

lime. (See section, p. 99.) In the uppermost of these divisions, but chiefly in the lowest, the remains of fish have been found, of the genus named by M. Agassiz, Cephalaspis, or buckler-headed, from the extraordinary shield which covers the head, and which has often been mistaken for that of a trilobite, of the division Asaphus. (See Fig. 276. p. 459.)

Fig. 274.

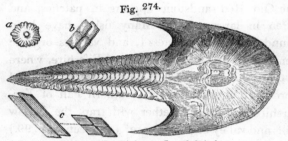

Cephlaspis Lyellii, Agass. Length 6¾ inches.

*This figure is from a specimen now in my collection, which I pro-
cured at Glammiss, in Forfarshire;* see other figures, Agassiz,
vol. ii. Tab. 1. *a.* & 1. *b.*

a, one of the peculiar scales with which the head is covered
when perfect. These scales are generally removed, as in the specimen above figured.

b, c, scales from different parts of the body and tail.

A gigantic species of fish of the genus Gyrolepis
has also been found by Dr. Fleming in the Old
Red sandstone of Fifeshire.*

* See Agassiz, Poissons Fossiles, tom. ii. p. 139.

CHAPTER XXII.

PRIMARY FOSSILIFEROUS STRATA.

Primary Fossiliferous or Transition Strata — Term " Grau-
wacké" — Silurian Group — Upper Silurian and Fossils —
Lower Silurian and Fossils — Trilobites — Graptolites —
Orthocerata — Occasional horizontality of Silurian Strata
— Cambrian Group — Endosiphonite.

WE have now arrived in the descending order at
those more ancient sedimentary rocks, which I
have called the Primary Fossiliferous (see p. 268.),
and to which Werner first gave the name of
Transition, for reasons fully explained and discussed in the 12th chapter. Many geologists have
also applied to these older strata the general name
of " grauwacké," by which the German miners
designate a variety of quartzose sandstone, which
is usually an aggregate of small fragments of
quartz, flinty-slate (or Lydian stone), and clay-
slate, cemented together by argillaceous matter.
But far too much importance has been attached to
this kind of rock, as if it were peculiar to a certain
epoch in the earth's history, whereas a similar
sandstone or grit is not only found sometimes in
the Old Red, and in the millstone grit of the
Coal, and in certain cretaceous formations of the
Alps — but even in some tertiary deposits.

In England, the Old Red sandstone has been

generally regarded as the base of the secondary series; but by some writers on the Continent, the Old Red and Coal formations have been classed as the upper members of the Transition series, a method adopted by Dr. Buckland, in his late Bridgewater Treatise. This classification, however, requires us to draw a strong line of demarcation between the Coal and the lower New Red sandstone group, which now that the fossils of these two groups are ascertained to be very analogous, becomes a more arbitrary division than that which separates the Old Red from the uppermost of the primary fossiliferous strata.

Professor Sedgwick and Mr. Murchison have lately proposed to subdivide all the English sedimentary strata below the Old Red sandstone into two leading groups, the upper of which may be termed the Silurian, and the inferior the Cambrian system. Mr. Murchison has applied the name of Silurian to the newer group, because these rocks may be best studied in that part of England and Wales which was included in the ancient British kingdom of the Silures. He has also formed four subdivisions of the Silurian system, which he has designated as the Ludlow, Wenlock, Caradoc, and Llandeilo, indicating thereby the places where the prevailing characters of each formation are most perfectly exhibited. The following Table explains the succession of these deposits.*

* See Murchison's Silurian System.

UPPER SILURIAN ROCKS.

	Prevailing Lithological Characters.	Thickness.	Organic Remains.
1. Ludlow formation.	**Upper Ludlow.** Micaceous grey sandstone. **Aymestry limestone.** Argillaceous limestone. **Lower Ludlow.** Shale, with concretions of limestone.	2000 Feet.	Mollusca marine, of almost every order, the Brachiopoda most abundant. Serpula, Corals, Sauroid fish, Fuci.
2. Wenlock formation.	**Wenlock limestone.** Concretionary limestone. **Wenlock shale.** Argillaceous shale.	1800 Feet.	Marine mollusca of various orders as before. Crustaceans of the Trilobite family. No vertebrata or plants.

LOWER SILURIAN ROCKS.

	Prevailing Lithological Characters.	Thickness.	Organic Remains.
3. Caradoc formation.	**Caradoc sandstones.** Flags of shelly limestone and sandstone, thick bedded white freestone.	2500 Feet.	Crinoidea, Corals, Mollusca, chiefly Brachiopoda, Trilobites.
4. Llandeilo formation.	**Llandeilo flags.** Dark coloured calcareous flags.	1200 Feet.	Mollusca, Trilobites.

x

UPPER SILURIAN ROCKS.

Ludlow formation. — This member of the upper Silurian group, as will be seen by the above table, is of great thickness, and subdivided into three parts.   Each of these may be distinguished near the town of Ludlow, and at other places in Shropshire and Herefordshire, by peculiar organic remains. The most remarkable fossils are the scales, ichthyodorulites, jaws, teeth, and coprolites of fish, of the upper Ludlow rock.*   As they are the oldest remains of vertebrated animals yet known to geologists, it is worthy of notice that they belong to fish of a high or very perfect organization.

Among the fossil shells are species of leptæna, orthis, terebratula, avicula, trochus, orthoceras, bellerophon, and others.†

Fig. 275.

Terebratula Wilsoni, Sow.   Ludlow formation.

Several species also of trilobite, an extinct species of crustacean, characteristic of the Silurian period in general, are found in the lower Ludlow limestone.   Those represented in the annexed figures, *Calymene Blumenbachii* and *Asaphus cau-*

* Murchison, Silurian System, p. 198, 199.   † Ibid.

datus, are common to this limestone, and to the Wenlock formation which succeeds next in the descending order.

Fig. 276.

Calymene Blumenbachii,
Brongniart, commonly called
" Dudley trilobite."

Fig. 277.

Asaphus caudatus.

Some of the Upper Ludlow sandstones are ripple-marked, thus affording evidence of gradual deposition; and the same may be said of the fine argillaceous shales of the Ludlow formation, which are of great thickness, and have been provincially named " mudstones," from their tendency to dissolve into mud. In these shales many zoophytes are found enveloped in an erect position, having evidently become fossil on the spots where they grew at the bottom of the sea. Among others, the graptolite is abundant. (See p. 462.) The facility with which these upper Silurian shales, when exposed to the weather, are resolved into mud, proves that, notwithstanding their antiquity, they are nearly in the state in which they were first thrown down at the bottom of the sea. All rocks, therefore, of the transition era of Werner, were not originally precipitated in a semi-crystalline state as was formerly pretended. (See p. 260.)

Wenlock formation. — The well-known rock of Dudley, so rich in organic remains, belongs to this member of the upper Silurian group, which consists in its higher division of limestone more or less crystalline, and highly charged with corals and encrinites of species distinct from those of the mountain limestone. In its lower part it is principally composed of argillaceous shale. In the Wenlock limestone, the chain-coral, called *Catenipora escharoides*, abounds. Among the shells appear the genera euomphalus, productus, atrypa, and many others.

Fig. 278.

Catenipora escharoides.

LOWER SILURIAN ROCKS.

Caradoc sandstone. — This formation, which is 2500 feet thick, consists chiefly of sandstones of various colours, with some subordinate beds of calcareous matter. Almost all the more abundant fossils belong to the same genera as those of the upper Silurian rocks, but the species are distinct.

Llandeilo formation. — This division, forming the base of the Silurian system, consists of hard dark-coloured flags, sometimes slightly micaceous, frequently calcareous, and especially distinguished

Fig. 279.

Asaphus Buchii,
Brong.

by containing the large trilobites *Asaphus Buchii* (Fig.
279.), and *A. tyranus.** There
are also several genera of mollusca in this deposit, and it
is an interesting fact, that with
many extinct forms of testacea peculiar to the
lower Silurian rocks, such as orthoceras, penta-
merus, spirifer, and productus, others are associated belonging to genera still existing, as nautilus, turbo, buccinum, turritella, terebratula, and
orbicula.†

No land plants seem yet to have been discovered in strata which can be unequivocally demonstrated to belong to the Silurian period.

In Norway and Sweden the Silurian strata
extend over a wide area, and so much resemble
those of England in lithological character and
fossils, that they will probably be found to be
divisible into similar groups. They are composed
of large deposits of sandstone, which is sometimes
found at the base of the system, resting on gneiss
and calcareous rocks, with orthocerata and corals;
the chain-coral, (Fig. 278.) before mentioned, being
very conspicuous; also fine bituminous shales containing graptolites. (Fig. 280.)

These bodies are supposed by Dr. Beck, of Copenhagen, to be fossil zoophytes, related to the

* Murchison, Silurian System, p. 222.   † Ibid. p. 351.

Fig. 280.

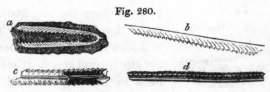

Graptolites, Linn.
a. b. Graptolites from Christiana, Norway.
c. d. Graptolites from the south of Sweden.

family of sea-pens, of which the living animals inhabit mud and slimy sediment.

In the limestones of Lake Michigan, in North America, and other regions bordering the great Canadian lakes, chain-corals and trilobites are also found, and from their fossils generally they seem to belong to the Silurian period. They contain certain orthocerata, which have a very peculiar structure. The siphuncle is very large, and has a tube running through its whole length, from the outside of which radii set off in verticilations extending to the inner wall of the syphon, these verticilations corresponding in number to the chambers of the shell. Mr. Stokes, who has described this division of orthoceratites, has formed them into a distinct genus, for which he has adopted the name of *Actinoceras*, proposed by Professor Bronn.* The actinoceras was not known as a British fossil, until lately discovered at Castle Espie, in the county of Down, in Ireland. (See Figs. 281, 282.)

* See Proceedings, Geol. Soc. 1838.

Fig. 281.

Actinoceras Simmsii, Stokes.
County of Down, Ireland.   Length of original 2 feet.

Fig. 282.

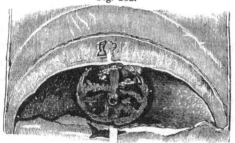

A. Simmsii, portion of the shell at *a*, Fig. 281., natural size, showing
the tube and its radii within the siphuncle.

Silurian strata occasionally horizontal.—The Silurian strata throughout a large part of the province of Skaraborg, in the south of Sweden, are
perfectly horizontal; the different subordinate
formations of sandstone, shale, and limestone,
occurring at corresponding heights in hills many
leagues distant from each other, with the same
mineral characters and organic remains. It is
clear that they have never been disturbed since
the time of their deposition, except by such gradual movements as those by which large areas in
Sweden and Greenland are now slowly and insensibly rising above or sinking below their former

level. The ancient limestone and shale also of
the Canadian lake district before mentioned, are
for the most part horizontal.

These facts are very important, as the more
ancient rocks are usually much disturbed, and
horizontality is a common character of newer
strata. Similar exceptions, however, occur in regard to the more modern or tertiary formations
which, in some places, as in the Alps, are not
only vertical, but in a reversed position. These
appearances accord best with the theory which
teaches that, at all periods, some parts of the
earth's crust have been convulsed by violent movements, which have been sometimes continued so
long, or so often repeated, that the derangement
has become excessive, while other spaces have escaped again and again, and have never once been
visited by the same kind of movement. Had
paroxysmal convulsions ever agitated simultaneously the entire crust of the earth, as some have
imagined, the primary fossiliferous strata would
nowhere have remained horizontal.

Cambrian Group.—Below the Silurian strata in
the region of the Cumberland lakes, in N. Wales,
Cornwall, and other parts of Great Britain, there
is a vast thickness of stratified rocks, for the most
part slaty, and devoid of fossils. In some few
places a few organic remains are detected specifically, and some of them generically, distinct

from those of the Silurian period. These rocks
have been called Cambrian by Professor Sedgwick,
because they are largely developed in N. Wales,
where they attain a thickness of several thousand
yards. They are chiefly formed of slaty sandstone and conglomerate, in the midst of which
is a limestone containing shells and corals, as at
Bala in Merionethshire. A slaty sandstone, forming the bottom of the Cambrian system in Snow-
don, contains shells of the family Brachiopoda, and
a few zoophytes. *

In some of the slate rocks of Cornwall, referred
by Professor Sedgwick to the Cambrian group,
cephalopoda of a very peculiar structure, called
Endosiphonites, have been detected, a form which
appears not yet to have been observed in the Silurian formation. The siphuncle in this shell is

Fig. 283.

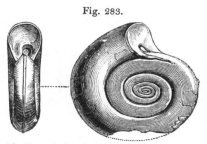

Endosiphonites carinatus, Ansted.†   Cambrian strata, Cornwall.

* Phillips's Geology, vol. i. p. 129. Lardner's Cyclop.
vol. xcvii.

† Camb. Phil. Trans. vol. vi. pl. 8. fig. 2.

ventral, in which character it differs both from ammonite, in which it is dorsal, and from nautilus, in which it is central, or nearly central.

Although the Cambrian group can scarcely yet be said to be established on the evidence of a distinct assemblage of fossils, yet so great is the thickness of strata beneath the lowest of the well-determined Silurian rocks, all of a date posterior to the creation of organic beings, that we may reasonably expect to be able to divide the primary fossiliferous strata into two groups.

CHAPTER XXIII.

ON THE DIFFERENT AGES OF THE VOLCANIC ROCKS.

Tests of relative age of volcanic rocks — Test by superposition and intrusion — By alteration of rocks in contact — Test by organic remains — Test of age by mineral character — Test by included fragments — Volcanic rocks of the Recent and Newer Pliocene periods — Miocene — Eocene — Cretaceous — Oolitic — New Red sandstone period — Carboniferous — Old Red sandstone period — Silurian — Upper and lower Cambrian periods — Relative ages of intrusive traps.

HAVING referred the sedimentary strata to a long succession of geological periods, we have next to consider how far the volcanic formations can be classed in a similar chronological order. The tests of relative age in this class of rocks are four : — 1st, superposition and intrusion, with or without alteration of the rocks in contact; 2d, organic remains; 3d, mineral character; 4th, included fragments of older rocks.

Test by superposition, &c. — If a volcanic rock rest upon an aqueous deposit, the former must be the newest of the two, but the like rule does not hold good where the aqueous formation rests upon the volcanic, for we have already seen (p. 181.)

that melted matter, rising from below, may penetrate a sedimentary mass without reaching the
surface, or may be forced in conformably between
two strata, as *b* at D in the annexed figure
(Fig. 284.), after which it may cool down and consolidate. Superposition, therefore, is not of the

Fig. 284.

same value as a test of age in the unstratified volcanic rocks as in fossiliferous formations. We can
only rely implicitly on this test where the volcanic
rocks are contemporaneous, not where they are
intrusive. Now they are said to be contemporaneous if produced by volcanic action, which was
going on simultaneously with the deposition of the
strata with which they are associated. Thus in
the section at D (Fig. 284.), we may perhaps ascertain that the trap *b* flowed over the fossiliferous
bed *c*, and that, after its consolidation, *a* was deposited upon it, *a* and *c* both belonging to the same
geological period. But if the stratum *a* be altered
by *b* at the point of contact, we must then conclude
the trap to have been intrusive, or if, in pursuing
b for some distance, we find at length that it cuts
through the stratum *a*, and then overlie it.

We may, however, be easily deceived in supposing a volcanic rock to be intrusive, when in
reality it is contemporaneous, for a sheet of lava,
as it spreads over the bottom of the sea, cannot
rest every where upon the same stratum, either
because these have been denuded, or because, if
newly thrown down, they thin out in certain places,
thus allowing the lava to cross their edges.   Besides, the heavy igneous fluid will often, as it moves
along, cut a channel into beds of soft mud and
sand. Suppose the submarine lava F, to have come
in contact in this manner with the strata *a, b, c,*
and that, after its consolidation, the strata *d, e,* are
thrown down in a nearly horizontal position, yet
so as to lie unconformably to F, the appearance of
subsequent intrusion

Fig. 285.

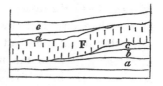

will here be complete,
although the trap is
in fact contempora-
neous. We must, un-
less we find the strata
d or *e* to have been altered at their junction, as if
by heat, not therefore hastily infer that the rock F
is intrusive.

When trap dikes were described in the 8th
chapter, they were shown to be more modern than
all the strata which they traverse. The ninety-
fathom dike in the Northumberland coal-field
(see section, Fig. 286.), passes through coal mea-

sures which are much disturbed.* The beds of

Fig. 286.

Magnesian limestone.

Coal.   Coal.

Dike.

Section in a quarry at Cullercoats, Northumberland.

overlying Magnesian limestone are not cut through
by the dike, but appear to be in the position in
which they were originally deposited in a hollow,
existing in the denuded surface formed by the
carboniferous strata and intrusive dike. Now here
the coal-measures were not only deposited, but had
been fissured before the fluid trap was introduced
to form the dike. It also appears by the truncated
edges of the Coal strata, and the abrupt termination
of the dike on which the Magnesian limestone
rests, that denudation had taken place at a period
intervening between the injection of the volcanic
matter and the deposition of the Magnesian limestone. Even in this case, however, although the
date of the volcanic eruption is brought within
narrow limits, it cannot be defined with precision ;
it may have happened either at the close of the
carboniferous period, or early in that of the lower
New Red sandstone, or between these two periods,

* See Mr. Winch's account, Geol. Trans. 1st ser. vol. iv.
p. 1.

when the state of the animate creation, and the physical geography of Europe were gradually changing from the type of the carboniferous era to that of the lower New Red formation.

In regard to all stratified volcanic tuffs, the test of age by superposition is strictly applicable to them, according to the already explained rules in the case of other sedimentary deposits. (See p. 272.)

Test of age by organic remains. — We have seen how, in the vicinity of active volcanos, scoriæ, pumice, fine sand, and fragments of rock are thrown up into the air, and then showered down upon the land, or into neighbouring lakes or seas. In the tuffs so formed, shells, corals, or any other durable organic bodies which may happen to be strewed over the bottom of a lake or sea will be imbedded in tuff, and thus continue as permanent memorials of the geological period when the volcanic eruption occurred. Tufaceous strata thus formed in the neighbourhood of Vesuvius, Etna, Stromboli, and other volcanos now active in islands or near the sea, may give information of the relative age of these tuffs at some remote future period when the fires of these mountains are extinguished. By such evidence we can distinctly establish the coincidence in age of volcanic rocks, and the different primary, secondary, and tertiary fossiliferous strata already considered.

The tuffs now alluded to are not exclusively
marine, but include, in some places, freshwater
shells, in others, the bones of terrestrial quadrupeds. The diversity of organic remains in
formations of this nature is perfectly intelligible,
if we reflect on the wide dispersion of ejected
matter during late eruptions, such as that of the
volcano of Coseguina, in the province of Nicaragua, January 19. 1835. Hot cinders and fine
scoriæ were then cast up to a vast height, and
covered the ground as they fell to the depth
of more than ten feet, and for a distance of
eight leagues from the crater in a southerly
direction. Birds, cattle, and wild animals were
scorched to death in great numbers, and buried in
these ashes. Some volcanic dust fell at Chiapa,
upwards of 1200 miles to windward of the volcano,
a striking proof of a counter current in the upper
region of the atmosphere, and some on Jamaica,
about 700 miles distant to the north-east. In the
sea also, at the distance of 1100 miles from the
point of eruption, Captain Eden of the Conway
sailed 40 miles through floating pumice, among
which were some pieces of considerable size.*

Test of age by mineral composition. — As sediment of homogeneous composition, when discharged
from the mouth of a large river, is often deposited

* Caldcleugh, Phil. Trans. 1836, p. 27., and Official Do-
cuments of Nicaragua.

simultaneously over a wide space, so a particular kind
of lava flowing from a crater, during one eruption,
may spread over an extensive area, as in Iceland
in 1783, when the melted matter, pouring from
Skaptar Jokul, flowed in streams in opposite directions, and caused a continuous mass, the extreme points of which were 90 miles distant from
each other.   This enormous current of lava varied
in thickness from 100 feet to 600 feet, and in
breadth, from that of a narrow river gorge to 15
miles.*   Now, if such a mass should afterwards be
divided into separate fragments by denudation, we
might still perhaps identify the detached portions
by their similarity in mineral composition. Nevertheless, this test will not always avail the geologist,
for, although there is usually a prevailing character in lava emitted during the same eruption,
and even in the successive currents flowing from
the same volcano, still, in many cases, the different
parts even of one lava-stream, or, as before stated,
of one continuous mass of trap, vary so much in
mineral composition and texture, as to render
these characters of minor importance when compared to their value in the chronology of the fossiliferous rocks.

It will, however, be seen in the description
which follows, of the European trap rocks of dif-

* See Principles, *Index,* " Skaptar Jokul."

ferent ages, that they had often a peculiar lithological character, resembling the differences before
remarked as existing between the modern lavas of
Vesuvius, Etna, and Chili. (See p. 158.)

It has been remarked that in Auvergne, the
Eifel, and other countries where trachyte and
basalt are both present, the trachytic rocks are for
the most part older than the basaltic. These
rocks do, indeed, sometimes alternate partially, as
in the volcano of Mont Dor, in Auvergne; but
the great mass of trachyte occupies in general an
inferior position, and is cut through and overflowed by basalt. It can by no means be inferred that trachyte predominated greatly at one
period of the earth's history and basalt at another,
for we know that trachytic lavas have been formed
at many successive periods, and are still emitted
from many active craters; but it seems that in
each region, where a long series of eruptions have
occurred, the more felspathic lavas have been first
emitted, and the escape of the more augitic
kinds has followed. The hypothesis suggested by
Mr. Scrope may, perhaps, afford a solution of
this problem. The minerals, he observes, which
abound in basalt are of greater specific gravity
than those composing the felspathic lavas; thus,
for example, hornblende, augite, and olivine, are
each more than three times the weight of water;
whereas common felspar, albite, and Labrador

felspar, have each scarcely more than $2\frac{1}{2}$ times the specific gravity of water; and the difference is increased in consequence of there being much more iron in a metallic state in basalt and greenstone than in trachyte and other felspathic lavas and traps. If, therefore, a large quantity of rock be melted up in the bowels of the earth by volcanic heat, the denser ingredients of the boiling fluid will sink to the bottom, and the lighter remaining above will be first propelled upwards to the surface by the expansive power of gases. Those materials, therefore, which occupied the lowest place in the subterranean reservoir will always be emitted last, and take the uppermost place on the exterior of the earth's crust.

Test by included fragments.—We may sometimes discover the relative age of two trap rocks, or of an aqueous deposit and the trap on which it rests, by finding fragments of one included in the other, in cases such as those before alluded to, where the evidence of superposition alone would be insufficient. It is also not uncommon to find conglomerates almost exclusively composed of rolled pebbles of trap, associated with stratified rocks in the neighbourhood of masses of intrusive trap. If the pebbles agree generally in mineral character with the latter, we are then enabled to determine the age of the intrusive rock by knowing that of the fossiliferous strata associated with the conglomerate. The origin of such conglomerates is explained by observing the shingle beaches composed of trap pebbles in modern volcanic islands, or at the base of Etna.

Recent and newer Pliocene period.—I shall now select examples of contemporaneous volcanic rocks of successive geological periods, that the reader may be convinced that the igneous causes have been in activity in all past ages of the world, and that they have been ever shifting the places where they have broken out at the earth's surface. One portion of the lavas, tuffs, and trap-dikes of Etna, Vesuvius, and the island of Ischia, have been produced within the historical era; another and a far more considerable part have originated at times immediately antecedent, when the waters of the Mediterranean were already inhabited by the existing species of testacea. The submarine foundations of Etna and Ischia have been upheaved to the great height of between 500 and 1500 feet above the level of the sea; and the same observations may be made respecting the base of many active volcanos which were first subaqueous vents, or, like Stromboli, half submerged, and then became subaerial, when the ancient bed of the sea was laid dry by elevation.

Older Pliocene period.—In Tuscany and the Campagna di Roma submarine volcanic tuffs are interstratified with the Older Pliocene strata of the Subapennine hills, in such a manner as to leave no doubt that they were the products of eruptions which occurred when the shelly marls and sands of the Subapennine hills were in the course of deposition.

Miocene period.— The most ancient volcanic rocks, consisting chiefly of trachyte, of the Upper and Lower Eifel, are intercalated between Miocene strata in such a manner, as to prove them to have been coeval in origin. The eruptions, however, of the same district were continued down to the Newer Pliocene era, or were at least renewed at that later period, so that showers of ashes from the Rhenish volcanos are interstratified with the loess, in which, we have already stated, shells of land and freshwater species occur identical with those now living in Europe.*

Eocene period.—The extinct volcanos of Auvergne and Cantal, in central France, commenced their eruptions in the Eocene period, but were most active during the Miocene era. In the lacustrine deposits, near those ancient volcanos, the lowest strata were evidently formed before any eruptions had occurred. They consist of sandstone and conglomerate, containing rounded pebbles of quartz, mica-schist, granite, and other hypogene rocks, composing the borders of the ancient lakes,

---

* See above, p. 297., and Principles, book iv.

but not the slightest intermixture of volcanic products can be detected. To these conglomerates succeeded argillaceous and calcareous marls containing Eocene shells, during the deposition of which some feeble signs of volcanic action began to show themselves. Above these, freshwater marls and limestones are seen frequently to alternate with volcanic tuff, and in them some fossils of the Miocene period are discovered. After the filling up or drainage of the ancient lakes, huge piles of trachytic and basaltic rocks, with volcanic breccias and conglomerates, accumulated to a thickness of several thousand feet, and were superimposed upon granite, or the contiguous lacustrine strata. The greater portion of these igneous rocks appear to have originated during the Miocene period, and extinct quadrupeds of that era, belonging to the genera Mastodon, Rhinoceros, and others, were buried in ashes and beds of alluvial sand and gravel, which owe their preservation to sheets of lava which spread over them.

Cretaceous period.—Although we have no proof of volcanic rocks erupted in England during the deposition of the chalk and green-sand, it must not be supposed that no theatres of igneous action existed in the cretaceous period. M. Virlet, in his account of the geology of the Morea, (p. 205.) has clearly shown that certain traps in Greece, called by him ophiolites, are of this date; as those, for example, which alternate conformably with cretaceous limestone and green-sand between Kastri and Damala in the Morea. They consist in great part of diallage rocks and serpentine, and of an amygdaloid with calcareous kernels, and a base of serpentine.

In certain parts of the Morea, the age of these volcanic rocks is established by the following proofs: first, the lithographic limestones (see p. 342.) of the cretaceous era are cut through by trap, and then a conglomerate occurs, at Nauplia and other places, containing in its calcareous cement many well-known fossils of the chalk and green-sand, together with pebbles formed of rolled pieces of the same ophiolite, which appear in the dikes above alluded to.

It was before stated that at Tercis, near Dax, in the department of the Landes, in the south of France, highly inclined strata of limestone and marl occur, containing the fossils of the chalk, the inclined strata being in great part concealed by unconformable tertiary formations. In one section in this district I observed, alternating with thin layers of volcanic tuff, vertical cretaceous beds, which are perfectly conformable. Such tuffs were probably the product of submarine eruptions in the cretaceous sea.

The traps of this country and of the neighbouring Pyrenees are generally ophitic, and many

Fig. 287.   Adour R.   Luy R.   Puy Arzet,

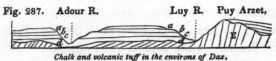

Chalk and volcanic tuff in the environs of Dax.
ɪ. Inclined beds of chalk and conformable volcanic tuff.
a. b. c. d. Gravel, sand, and tertiary strata.

French geologists conceive them to be newer than the cretaceous period, and therefore tertiary; but I know of no sections which demonstrate this point. M. Charpentier has argued that the ophites of the Pyrenees were more modern than all the secondary strata of that chain, because in the conglomerates constituting the upper part of the cretaceous series on the flanks of the Pyrenees, no rolled pebbles of ophite have been found.[*] But this negative fact may be explained by supposing that, in the cretaceous sea, which occupied the space where the Pyrenees now stand, the ophitic eruptions were submarine, and never formed islands or shoals exposed to denudation.

The age of the trap of Antrim in Ireland, before described, as altering the chalk by its dikes (p. 176.), is uncertain. It is newer than the chalk of that region, which it cuts through and overflows; and, perhaps, it belongs to some one of the tertiary periods. As wood-coal and coniferous fossil trees have been found associated with it on the eastern shores of Lough Neagh, these plants

[*] Charpentier, Essai Geog. sur les Pyrénées, p. 524.

may hereafter throw light on this chronological question.*

Period of Oolite and Lias.—Although the green and serpentinous trap rocks of the Morea belong chiefly to the cretaceous era, as before mentioned, yet it seems that some eruptions of similar rocks began during the oolitic period †; and it is probable, that a large part of the trappean masses, called ophiolites in the Apennines, and associated with the limestone of that chain, are of corresponding age.

Whether part of the volcanic rocks of the Hebrides, in our own country, originated contemporaneously with the lias and oolite which they traverse and overlie, remains to be ascertained.

Trap of the New Red sandstone period. — In the southern part of Devonshire, trappean rocks are associated with new red sandstone, and, according to Mr. De la Beche, have not been intruded subsequently into the sandstone, but were produced by contemporaneous volcanic action. Some beds of grit, mingled with ordinary red marl, resemble sands ejected from a crater; and in the stratified conglomerates occurring near Tiverton are many angular fragments of trap porphyry, some of them one or two tons in weight, intermingled with

* Dr. Berger, Geol. Trans. 1st series, vol. iii. p. 168.
† Boblaye and Virlet, Morea, p. 23.

pebbles of other rocks. These angular fragments were probably thrown out from volcanic vents, and fell upon sedimentary matter then in the course of deposition.*

Carboniferous period. — Two classes of contemporaneous trap rocks have been ascertained by Dr. Fleming to occur in the coal-field of the Forth in Scotland. The newest of these, connected with the higher series of coal-measures, is well exhibited along the shores of the Forth, in Fifeshire, where they consist of basalt with olivine, amygdaloid, greenstone, wacké, and tuff. They appear to have been erupted while the sedimentary strata were in a horizontal position, and to have suffered the same dislocations which those strata have subsequently undergone. In the volcanic tuffs of this age are found not only fragments of limestone, shale, flinty slate, and sandstone, but also pieces of coal.

The other or older class of carboniferous traps are traced along the south margin of Stratheden, and constitute a ridge parallel with the Ochils, and extending from Stirling to near St. Andrew's. They consist almost exclusively of greenstone, becoming, in a few instances, earthy and amygdaloidal. They are regularly interstratified with the sandstone, shale, and ironstone of the lower coal-

* De la Beche, Geol. Proceedings, No. 41. p. 196.

measures, and, on the East Lomond, with Mountain limestone.*

Trap of the Old Red sandstone period. — By referring to the section explanatory of the structure of Forfarshire, already given (p. 99.), the reader will perceive that beds of conglomerate, No. 3., occur in the middle of the old red sandstone system, 1, 2, 3, 4. The pebbles in these conglomerates are sometimes composed of granitic and quartz rocks, sometimes exclusively of different varieties of trap, which, although purposely omitted in the above section, are often found, either intruding themselves in amorphous masses and dikes into the older fossiliferous tilestones, No. 4., or alternating with them in conformable beds. All the different divisions of the red sandstone, 1, 2, 3, 4, are occasionally intersected by dikes, but they are very rare in Nos. 1. and 2., the upper members of the group consisting of red shale and red sandstone. These phenomena, which occur at the foot of the Grampians, are repeated in the Sidlaw Hills; and it appears that in this part of Scotland volcanic eruptions were most frequent in the earlier part of the old red sandstone period.

The trap rocks alluded to consist chiefly of felspathic porphyry and amygdaloid, the kernels of the latter being sometimes calcareous, often

* Fleming MS.   Part of this tract I have myself examined with Dr. F.

chalcedonic, and forming beautiful agates. We meet also with claystone, clinkstone, greenstone, compact felspar, and tuff. Some of these rocks flowed as lavas over the bottom of the sea, and enveloped quartz pebbles which were lying there, so as to form conglomerates with a base of greenstone, as is seen in Lumley Den, in the Sidlaw Hills. On either side of the axis of this chain of hills (see section, p. 99.), the beds of massive trap, and the tuffs composed of volcanic sand and ashes, dip regularly to the south-east or north-west, conformably with the shales and sandstones.

Dr. Fleming has observed similar trap rocks in the old red sandstone of northern Fifeshire, where they are covered immediately by the yellow sandstone which forms the base of the mountain limestone and coal-measures.

Silurian period. — It appears from the investigations of Mr. Murchison in Shropshire, that when the lower Silurian strata of that county were accumulating, there were frequent volcanic eruptions beneath the sea; and the ashes and scoriæ then ejected gave rise to a peculiar kind of tufaceous sandstone or grit, dissimilar to the other rocks of the Silurian series, and only observable in places where syenitic and other trap rocks protrude.* These tuffs occur on the flanks of the Wrekin and

* Murchison, Silurian System, &c. p. 230.

Caer Caradoc, and contain Silurian fossils, such as casts of encrinites, trilobites, and mollusca. Although fossiliferous, the stone resembles a sandy claystone of the trap family.*

Thin layers of trap, only a few inches thick, alternate, in some parts of Shropshire and Montgomeryshire, with sedimentary strata of the lower Silurian system. This trap consists of slaty porphyry and granular felspar rock, the beds being traversed by joints like those in the associated sandstone, limestone, and shale, and having the same strike and dip. †

In Radnorshire, there is an example of twelve bands of stratified trap alternating with Silurian schists and flagstones in a thickness of 350 feet. The bedded traps consist of felspar-porphyry, clinkstone, and other varieties; and the interposed Llandeilo flags are of sandstone and shale, with trilobites and graptolites. ‡

Cambrian volcanic rocks. — In Pembrokeshire stratified greenstone, felspar-rock, and a breccia containing fragments of trap, alternate conformably in thick parallel masses with regularly stratified sandstone and schist of the *upper Cambrian* system. These trappean masses, says Mr. Murchison, must have been evolved at intervals from volcanic fissures at the bottom of the

* Murchison, Silurian System, &c. p. 230.
† Ibid. p. 272.   ‡ Ibid. p. 325.

sea, when the sand, pebbles, and mud, now form‑
ing the accompanying sedimentary rocks, were
deposited.*

Professor Sedgwick, in his account of the geo‑
logy of Cumberland, has described various trap‑
rocks which accompany the green slates of the
Cambrian system, beneath a limestone containing
organic remains. Different felspathic and por‑
phyritic rocks and greenstones occur, not only in
dikes, but in conformable beds; and there is occa‑
sionally a passage from these igneous rocks to some
of the green quartzose slates. Professor Sedg‑
wick supposes these porphyries to have originated
contemporaneously with the stratified chloritic
slates, the materials of the slates having been sup‑
plied, in part at least, by submarine eruptions
oftentimes repeated.†

* Murchison, Silurian System, &c. p. 404.
† Geol. Trans. 2d series, vol. iv. p. 55.

CHAPTER XXIV.

ON THE DIFFERENT AGES OF THE PLUTONIC ROCKS.

Difficulty in ascertaining the precise age of a plutonic rock
— Test of age by relative position — Test by intrusion
and alteration — Test by mineral composition — Test by
included fragments — Recent and Pliocene plutonic rocks,
why invisible — Tertiary plutonic rocks in the Andes —
Granite altering Cretaceous rocks — Granite altering Lias
in the Alps and in Sky — Granite of Dartmoor altering
Carboniferous strata — Granite of the Old Red sandstone
period — Syenite altering Silurian strata in Norway —
Blending of the same with gneiss — Most ancient plutonic
rocks — Granite protruded in a solid form — On the probable age of the granite of Arran, in Scotland.

WHEN we adopt the igneous theory of granite, as
explained in the 9th chapter, and believe that
different plutonic rocks have originated at successive periods beneath the surface of the planet, we
must be prepared to encounter greater difficulty
in ascertaining the precise age of such rocks, than
in the case of volcanic and fossiliferous formations.
We must bear in mind, that the evidence of the
age of each contemporaneous volcanic rock was derived, either from lavas poured out upon the ancient
surface, whether in the sea or in the atmosphere,
or from tuffs and conglomerates, also deposited at
the surface, and either containing organic remains

themselves, or intercalated between strata containing fossils.   But all these tests fail when we endeavour to fix the chronology of a rock, which has
crystallized from a state of fusion in the bowels of
the earth.   In that case, we are reduced to the
following tests; 1st, relative position ; 2dly, intrusion, and alteration of the rocks in contact; 3dly,
mineral characters; 4thly, included fragments.

  Test of age by relative position.—Unaltered fossiliferous strata of every age are met with reposing
immediately on plutonic rocks ; as at Christiania, in
Norway, where the Newer Pliocene deposits rest
on granite ; in Auvergne, where the freshwater
Eocene strata, and at Heidelberg, on the Rhine,
where the New Red sandstone, occupy a similar
place.   In all these, and similar instances, inferior-
ity in position is connected with the superior antiquity of granite.   The crystalline rock was solid
before the sedimentary beds were superimposed,
and the latter usually contain in them rounded
pebbles of the subjacent granite.

  Test by intrusion and alteration. — But when
plutonic rocks send veins into strata, and alter
them near the point of contact, in the manner before described (p. 204.), it is clear that, like intrusive traps, they are newer than the strata which
they invade and alter.   Examples of the application of this test will be given in the sequel.

  Test by mineral composition.—Notwithstanding a

general uniformity in the aspect of plutonic rocks, we have seen in the 9th chapter that there are many varieties, such as Syenite, Talcose granite, and others. One of these varieties is sometimes found exclusively prevailing throughout an extensive region, where it preserves a homogeneous character; so that having ascertained its relative age in one place, we can easily recognize its identity in others, and thus determine from a single section the chronological relations of large mountain masses. Having observed, for example, that the syenitic granite of Norway, in which the mineral called zircon abounds, has altered the Silurian strata wherever it is in contact, we do not hesitate to refer all masses of the same zircon-syenite in the south of Norway to the same era. (See p. 242.)

Some have imagined that the age of different granites might, to a great extent, be determined by their mineral characters alone; syenite, for instance, or granite with hornblende, being more modern than common or micaceous granite. But modern investigations have proved these generalizations to have been premature. The syenitic granite of Norway already alluded to may be of the same age as the Silurian strata, which it traverses and alters, or may belong to the Old Red sandstone period; whereas the granite of Dartmoor, although consisting of mica, quartz, and felspar, is newer than the Coal.   (See p. 499.)

Test by included fragments. — This criterion can rarely be of much importance, because the fragments involved in granite are usually so much altered, that they cannot be referred with certainty to the rocks whence they were derived.   In the White Mountains in North America, according to Professor Hubbard, a granite vein traversing granite, contains fragments of slate and trap, which must have fallen into the fissure when the fused materials of the vein were injected from below*, and thus the granite is shown to be newer than certain superficial slaty and trappean formations.

Recent and Pliocene plutonic rocks, why invisible. — The explanation already given in the 8th and 9th chapters of the probable relation of the plutonic to the volcanic formations, will naturally lead the reader to infer, that rocks of the one class can never be produced at or near the surface without some members of the other being formed below simultaneously, or soon afterwards.   It is not uncommon for lava streams to require more than ten years to cool in the open air; and where they are of great depth, a much longer period.   The melted matter poured from Jorullo, in Mexico, in the year 1759, which accumulated in some places to

* Silliman's Journ. No. 69. p. 123.

the height of 550 feet, was found to retain a high temperature half a century after the eruption.[*] We may conceive, therefore, that great masses of subterranean lava may remain in a red-hot or incandescent state in the volcanic foci for immense periods, and the process of refrigeration may be extremely gradual. Sometimes, indeed, this process may be retarded for an indefinite period, by the accession of fresh supplies of heat; for we find that the lava in the crater of Stromboli, one of the Lipari islands, has been in a state of constant ebullition for the last two thousand years; and we must suppose this fluid mass to communicate with some caldron or reservoir of fused matter below. In the Isle of Bourbon, also, where there has been an emission of lava once in every two years for a long period, the lava below can scarcely fail to have been permanently in a state of liquefaction. If then it be a reasonable conjecture, that about 2000 volcanic eruptions occur in the course of every century, either above the waters of the sea or beneath them[†], it will follow, that the quantity of plutonic rock generated, or in progress during the Recent epoch, must already have been considerable.

But as the plutonic rocks originate at some depth in the earth's crust, they can only be ren-

[*] See Principles, Index, " Jorullo."
[†] Ibid. Index, " Volcanic Eruptions."

dered accessible to human observation by subsequent upheaval and denudation. Between the
period when a plutonic rock crystallizes in the
subterranean regions, and the era of its protrusion
at any single point of the surface, one or two
geological periods must usually intervene. Hence,
we must not expect to find the Recent or Newer
Pliocene granites laid open to view, unless we are
prepared to assume that sufficient time has elapsed
since the commencement of the Newer Pliocene
period for great upheaval and denudation. A
plutonic rock, therefore, must, in general, be of
considerable antiquity relatively to the fossiliferous
and volcanic formations, before it becomes extensively visible. As we know that the upheaval of land
has been sometimes accompanied in South America
by volcanic eruptions and the emission of lava, we
may conceive the more ancient plutonic rocks to
be forced upwards to the surface by the newer
rocks of the same class formed successively below,
—subterposition in the plutonic, like superposition in the sedimentary rocks, being usually characteristic of a newer origin.

In the accompanying diagram, Fig. 288., an attempt is made to show the inverted order in which
sedimentary and plutonic formations may occur in
the earth's crust.

Fig. 288.

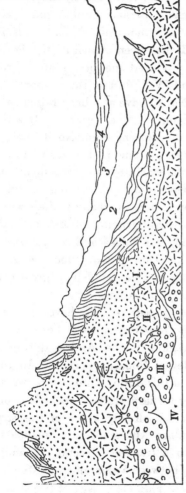

Diagram showing the relative position which the plutonic and sedimentary formations of different ages may occupy.

I.   Primary plutonic.
II. Secondary plutonic.
III. Tertiary plutonic.
IV. Recent plutonic.

4. Recent strata.
3. Tertiary strata.
2. Secondary strata.
1. Primary fossiliferous strata.

The metamorphic rocks are not indicated in this diagram, but the student may learn from the Frontispiece what place they would occupy if portions of the strata Nos. 1. and 2., invaded by granite, had become metamorphic.

The oldest plutonic rock, No. I., has been up-heaved at successive periods until it has become exposed to view in a mountain-chain. This pro-trusion of No. I. has been caused by the igneous agency which produced the new plutonic rocks Nos. II. III. and IV. Part of the primary fossil-iferous strata, No. 1., have also been raised to the surface by the same gradual process. It will be observed that the Recent *strata* No. 4., and the Recent *granite* or plutonic rock No. IV., are the most remote from each other in position, although of contemporaneous date. According to this hy-pothesis, the convulsions of many periods will be required before *Recent* granite will be upraised so as to form the highest ridges and central axes of mountain-chains. During that time the *Recent* strata No. 4. might be covered by a great many newer sedimentary formations.

Tertiary plutonic rocks. — We have seen that great upheaving movements have been experi-enced in the region of the Andes, during the Recent and Newer Pliocene periods. In some part, therefore, of this chain, if any where, we may hope to discover tertiary plutonic rocks laid open to view. What we already know of the struc-ture of the Chilian Andes seems to realize this expectation. In a transverse section, examined by Mr. Darwin, between Valparaiso and Mendoza, the Cordillera was found to consist of two separate and parallel chains, formed of sedimentary rocks of different ages, the strata in both resting on plutonic rocks, by which they have been altered. In the western or oldest range, called the Peuquenes, are black calcareous clay-slates, rising to the height of nearly 14,000 feet above the sea, in which are shells of the genera Gryphæa, Turritella, Terebratula, and Ammonite. These rocks are supposed to be of the age of the central parts of the secondary series of Europe. They are penetrated and altered by dikes and mountain masses of a plutonic rock, which has the texture of ordinary granite, but rarely contains quartz, being a compound of albite and hornblende.

The second or eastern chain consists chiefly of sandstones and conglomerates, of vast thickness, the materials of which are derived from the ruins of the western chain. The pebbles of the conglomerates are, for the most part, rounded fragments of the fossiliferous slates before mentioned. The resemblance of the whole series to certain tertiary deposits on the shores of the Pacific, not only in mineral character, but in the imbedded lignite and silicified wood, leads to the conjecture that they also are tertiary. Yet these strata are not only associated with trap rocks and volcanic tuffs, but are also altered by a granite newer than that of the western chain, and consisting of quartz, felspar, and talc. They are traversed, moreover, by dikes of the same granite, and by numerous veins of iron, copper, arsenic, silver, and gold; all of which can be traced to the underlying granite.* We have, therefore, strong ground to presume that the plutonic rock, here exposed on a large scale in the Chilian Andes, is of later date than certain tertiary formations.

Cretaceous period. — It was stated (p. 245.) that chalk as well as lias have been altered by granite in the eastern Pyrenees. Whether such granite be cretaceous or tertiary cannot easily be decided.

Fig. 289.

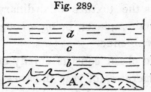

Suppose *b*, *c*, *d*, to be three members of the Cretaceous series, the lowest of which, *b*, has been altered by the granite A, the modifying influence not having extended so far as *c*, or having but slightly affected its lowest beds. Now it can rarely be possible for the geologist to decide whether the beds *d* existed at the time of the intrusion of A, and alteration of *b* and *c*, or whether they were subsequently thrown down upon *c*.

As some Cretaceous rocks, however, have been raised to the height of more than 9000 feet in the Pyrenees, we must not assume that plutonic formations of the same age may not have been

* Darwin, pp. 390. 406.

brought up and exposed by denudation, at the
height of 2000 or 3000 feet on the flanks of that
chain.

Period of Oolite and Lias. — In the department
of the Hautes Alpes, in France, near Vizille, M.
Elie de Beaumont traced a black argillaceous
limestone, charged with belemnites, to within a few
yards of a mass of granite.   Here the limestone

Fig. 290.

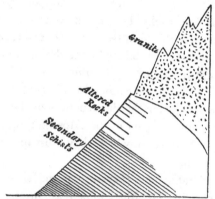

Junction of granite with Jurassic or Oolite strata in the Alps, near Champoleon.

begins to put on a granular texture, but is extremely fine-grained.   When nearer the junction
it becomes grey, and has a saccharoid structure.
In another locality, near Champoleon, a granite
composed of quartz, black mica, and rose-coloured
felspar, is observed partly to overlie the secondary
rocks, producing an alteration which extends for

about thirty feet downwards, diminishing in the beds which lie farthest from the granite. (See Fig. 290.) In the altered mass the argillaceous beds are hardened, the limestone is saccharoid, the grits quartzose, and in the midst of them is a thin layer of an imperfect granite. It is also an important circumstance, that near the point of contact, both the granite and the secondary rocks become metalliferous, and contain nests and small veins of blende, galena, iron, and copper pyrites. The stratified rocks become harder and more crystalline, but the granite, on the contrary, softer and less perfectly crystallized near the junction.*

Although the granite is incumbent in the above section (Fig. 290.), we cannot assume that it overflowed the strata, for the disturbances of the rocks are so great in this part of the Alps that they seldom retain the position which they must originally have occupied.

A considerable mass of syenite, in the Isle of Sky, is described by Dr. MacCulloch as intersecting limestone and shale, which are of the age of the lias. † The limestone, which, at a greater distance from the granite, contains shells, exhibits no traces

* Elie de Beaumont, sur les Montagnes de l'Oisans, &c., Mém. de la Soc. d'Hist. Nat. de Paris, tome v.

† See Murchison, Geol. Trans., 2nd series, vol. ii. part ii. pp. 311—321.

of them near its junction, where it has been converted into a pure crystalline marble.*

At Predazzo, in the Tyrol, secondary strata, some of which are limestones of the Oolitic period, have been traversed and altered by plutonic rocks, one portion of which is an augitic porphyry, which passes insensibly into granite. The limestone is changed into granular marble, with a band of serpentine at the junction.†

Carboniferous period. — The granite of Dartmoor, in Devonshire, was formerly supposed to be one of the most ancient of the plutonic rocks, but is now ascertained to be posterior in date to the culm-measures of that county, which, from their position, and as containing true coal-plants, are regarded by Professor Sedgwick and Mr. Murchison as members of the true carboniferous series. This granite, like the syenitic granite of Christiania, has broken through the stratified formations without much changing their strike. Hence, on the north-west side of Dartmoor, the successive members of the culm-measures abut against the granite, and become metamorphic as they approach. These strata are also penetrated by granite veins and plutonic dikes, called "elvans."‡

* Western Islands, vol. i. p. 330. plate 18. figs. 3, 4.
† Von Buch, Annales de Chimie, &c.
‡ Proceedings of Geol. Soc., vol. ii. p. 562.

The granite of Cornwall is probably of the same date, and, therefore, as modern as the Carboniferous strata, if not much newer.

Old Red sandstone period. — The plutonic rocks of the Malvern hills, in Worcestershire, consist of a granitic compound of quartz, felspar, and hornblende, or occasionally of quartz, mica, and felspar, which passes into syenite and greenstone.* This rock has altered the adjacent Silurian strata into well characterized metamorphic schists, principally chloritic and micaceous-schist, with some gneiss, and has dislocated and reversed the position of the beds of the Silurian and Old Red sandstone. There are indications, says Mr. Murchison, of several periods of movement, by which the strata were forced up and folded back, but the chief outburst was after the accumulation of the Silurian and part of the Old Red system, and anterior to the formation of the coal-beds, which are undisturbed. †

Silurian period. — I have already alluded to the granite near Christiania, in Norway, as being newer than the Silurian strata of that region. Its posteriority in date to limestones containing orthocerata and trilobites, has long been celebrated, it being twenty-five years since Von Buch first announced the discovery. The proofs consist

* Mr. L. Horner, Geol. Trans., 1st ser., vol. i. p. 281.
† Silurian System, p. 425.

in the penetration of granite veins into the shale
and limestone, and the alteration of the strata, for
a considerable distance from the point of contact,
both of these veins and the central mass from
which they emanate.   (See p. 215.)   Von Buch
supposed that the plutonic rock alternated with
the fossiliferous strata, and that large masses of
granite were sometimes incumbent upon the strata;
but this idea was erroneous, and arose from the
fact that the beds of shale and limestone often dip
towards the granite up to the point of contact,
appearing as if they would pass under it in mass,
as at *a*, Fig. 291., and then again on the opposite
side of the same mountain, as at *b*, dip away from
the same granite.   When the junctions, however,

Fig. 291.

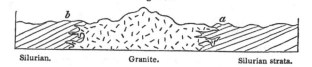

Silurian.   Granite.   Silurian strata.

are carefully examined, it is found that the plutonic rock intrudes itself in veins, and no where
covers the fossiliferous strata in large overlying
masses, as is so commonly the case with trappean
formations.*

Now this granite, which is more modern than

* See the Gæa Norvegica and other works of Keilhau,
with whom I examined this country.

the Silurian strata of Norway, also sends veins in
the same country into an ancient formation of
gneiss; and the relations of the plutonic rock and
the gneiss, at their junction, are full of interest
when we duly consider the wide difference of
epoch which must have separated their origin.

The length of this interval of time is attested
by the following facts: — The fossiliferous, or
transition beds, rest unconformably upon the
truncated edges of the gneiss, the inclined strata
of which had been disturbed and denuded before the sedimentary beds were superimposed.
(See Fig. 292.)

Fig. 292.

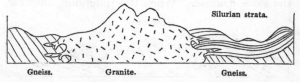

Gneiss.   Granite.   Gneiss.

Granite sending veins into Silurian strata and Gneiss, — Christiania, Norway.

The signs of denudation are twofold; first, the
surface of the gneiss is seen occasionally on the
removal of the newer beds, containing organic
remains, to be scored and polished; secondly,
pebbles of gneiss have been found in some of the
transition strata. Between the origin, therefore,
of the gneiss and the granite there intervened,
first, the period when the strata of gneiss were
inclined; secondly, the period when they were

denuded; thirdly, the period of the deposition of the transition deposits. Yet the granite produced, after this long interval, is often so intimately blended with the ancient gneiss, at the point of junction, that it is impossible to draw any other than an arbitrary line of separation between them; and where this is not the case, tortuous veins of granite pass freely through gneiss, ending sometimes in threads, as if the older rock had offered no resistance to their passage. It seems necessary, therefore, to conceive that the gneiss was softened and more or less melted when penetrated by the granite. But had such junctions alone been visible, and had we not learnt, from other sections, how long a period elapsed between the consolidation of the gneiss and the injection of this granite, we might have suspected that the gneiss was scarcely solidified, or had not yet assumed its complete metamorphic character, when invaded by the plutonic rock. From this example we may learn how impossible it is to conjecture whether certain granites in Scotland, and other countries, which send veins into gneiss and other metamorphic rocks, are primary, or whether they may not belong to some secondary or tertiary period.

Most ancient granites. — It is not half a century since the doctrine was very general that all granitic rocks were *primitive*, that is to say, that they originated before the deposition of the first sedimen-

tary strata, and before the creation of organic
beings. (See p. 20.) But so greatly are our views
now changed, that we find it no easy task to point
out a single mass of granite demonstrably more
ancient than all the known fossiliferous deposits.
Could we discover some Lower Cambrian strata
resting immediately on granite, there being no
alterations at the point of contact, nor any intersecting granitic veins, we might then affirm the
plutonic rock to have originated before the oldest
known fossiliferous strata. Still it would be pre-
sumptuous to suppose that when a small part only
of the globe has been investigated, we are acquainted with the oldest fossiliferous strata in the
crust of our planet. Even when these are found,
we cannot assume that there never were any antecedent strata containing organic remains, which may
have become metamorphic. If we find pebbles of
granite in a conglomerate of the Lower Cambrian
system, we may then feel assured that the parent
granite was formed before the Lower Cambrian
formation. But if the incumbent strata be merely
Silurian or Upper Cambrian, the fundamental
granite, although of high antiquity, may be posterior in date to *known* fossiliferous formations.

Protrusion of solid granite. — In part of Sutherlandshire, near Brora, common granite, composed
of felspar, quartz, and mica, is in immediate contact
with Oolitic strata, and has clearly been elevated to

the surface at a period subsequent to the deposition of those strata. * Professor Sedgwick and Mr. Murchison conceive that this granite has been upheaved in a solid form; and that in breaking through the submarine deposits, with which it was not perhaps originally in contact, it has fractured them so as to form a breccia along the line of junction.   This breccia consists of fragments of shale, sandstone, and limestone, with fossils of the oolite, all united together by a calcareous cement. The secondary strata, at some distance from the granite, are but slightly disturbed, but in proportion to their proximity the amount of dislocation becomes greater.

If we admit that solid hypogene rocks, whether stratified or unstratified, have in such cases been driven upwards so as to pierce through yielding sedimentary deposits, we shall be enabled to account for many geological appearances otherwise inexplicable.   Thus, for example, at Weinböhla and Hohnstein, near Meissen, in Saxony, a mass of granite has been observed covering strata of the cretaceous and oolitic periods for the space of between 300 and 400 yards square.   It appears clearly from a recent memoir of Dr. B. Cotta on this subject †, that the granite was thrust into its

* Murchison, Geol. Trans. 2d series, vol. ii. p. 307.
† Geognostische Wanderungen, Leipzig, 1838.

actual position when solid.   There are no intersecting veins at the junction — no alteration as if
by heat, but evident signs of rubbing, and a breccia in some places, in which pieces of granite are
mingled with broken fragments of the secondary
rocks.   As the granite overhangs both the lias and
chalk, so the lias is in some places bent over strata
of the cretaceous era.

Age of the granite of Arran. — In this island, the
largest in the Firth of Clyde, on the west coast of
Scotland, the four great classes of rocks, the fossiliferous, volcanic, plutonic, and metamorphic, are
all conspicuously displayed within a very small
area, and with their peculiar characters strongly
contrasted.   In the north of the island the granite
rises to the height of nearly 3000 feet above the
sea, terminating in mountainous peaks.   On the
flanks of the same mountains are chloritic-schists,
blue roofing-slate, and other rocks of the metamorphic order (A), into which the granite (B)
sends veins.   These schists are highly inclined.
On their truncated edges rest unconformable beds
of conglomerate and sandstone, to which succeed
various shales and limestones, containing fossils of
the carboniferous period.   All these secondary
strata (c) are much tilted and inclined near the
hypogene rocks; but are horizontal at a distance
from them, as in the south of Arran.   Lastly, the
volcanic rocks (D), consisting of greenstone, pitch-

Fig. 293.   Section of Arran.

A, Crystalline, or metamorphic schist.   B, Granite.
C, Conglomerate, sandstone, limestone, and shale.   D, Trap.

stone, claystone, porphyry, and other varieties, traverse all the preceding formations, cutting through the granite in dikes (*d*), as well as through the sandstone; which last they also overlie in dense masses, from 50 to 700 feet in thickness.

Now as the different kinds of trap intersect all the other formations, they are certainly the newest rocks in Arran. The red sandstone and other secondary strata are older than the trap, but newer than the metamorphic schists, for the Red sandstone conglomerates not only rest unconformably upon the schists, but contain rounded pebbles of those crystalline strata. It is equally certain that the schists are the oldest rocks in the island: they are more ancient than the trap and red sandstone, for reasons already stated; and the granite must be of newer origin, because it penetrates them in veins. The only chronological point, therefore, in which there can be any ambiguity, relates to the plutonic formations. They are more modern, as before remarked, than the crystalline schists; but can we decide them to be likewise younger than the secondary sandstones?

Now it is a curious and most striking fact, that no pebbles of granite have ever been found in the conglomerates of the red sandstone in Arran, although careful search has been made for them by many geologists; and although puddingstones in general are chiefly made up of fragments of older rocks of the immediate vicinity. The total absence of such pebbles has justly been a theme of wonder to those who have visited Arran, and have seen that the conglomerates are several hundred feet in thickness, and that they occur at the base of the granite mountains, which tower above them in far bolder and more picturesque forms than those of similar composition in other parts of Scotland. We may at once infer, with confidence, that when the sandstone and conglomerate were formed, no granite had reached the surface, or had been exposed to denudation in this region : the crystalline schists were ground into sand and shingle when these puddingstones were accumulated, but the waves had never acted upon the granite, which sends its veins into the schist. Are we then to conclude, that the schists suffered denudation before they had been invaded by granite ? This opinion, although it cannot be disproved, is by no means fully borne out by the

evidence. At the time when the red sandstone
was formed, the metamorphic strata may have
formed islands in the sea, as in Fig. 294., over which
the breakers rolled, or from which torrents and
rivers descended, carrying down gravel and sand.
The plutonic rock (B) may have been previously

Fig. 294.

injected at a certain depth below, and yet may
never have been exposed to denudation.

As to the time and manner of the subsequent
protrusion of the hypogene rocks in Arran, these
are questions into which I have not space to enter
at present : I shall merely observe, that those crystalline rocks may have been thrust up bodily, in a
solid form ; and it is clear that, during or since
the period of their emergence, they have undergone great aqueous denudation. This action is
confirmed by three distinct kinds of proofs : 1st,
The occurrence of scattered pebbles and huge erratic blocks of granite and schist over the surface
of Arran and the adjacent mainland ; 2dly, The
abrupt truncation of dikes, such as those at *d*
(Fig. 293.), cut off on the surface of the granite ;
3dly, The fact, that not only the secondary
strata but the enormous masses of trap which ac-

company and overlie them, terminate suddenly on
reaching the borders of the granite and schist,
towards which they often present a steep escarpment, and over which, for some distance at least,
they must originally have extended.*

* In the works of Drs. Hutton and MacCulloch, and in
the Memoirs of Messrs. Von Dechen and Oeynhausen, and
that of Professor Sedgwick and Mr. Murchison (Geol. Trans.
2d series) and others, whose observations I have verified
on the spot, the reader will find a full description of the
geology of Arran.

CHAPTER XXV.

ON THE DIFFERENT AGES OF THE METAMORPHIC ROCKS.

Age of each set of metamorphic strata twofold — Test of age by fossils and mineral character not available — Test by superposition ambiguous — Conversion of dense masses of fossiliferous strata into metamorphic rocks — Limestone and shale of Carrara — Metamorphic strata of modern periods in the Alps of Switzerland and Savoy — Why the visible crystalline strata are none of them very modern — Order of succession in metamorphic rocks — Uniformity of mineral character — Why the metamorphic strata are less calcareous than the fossiliferous.

ACCORDING to the theory adopted in the 11th chapter, the age of each set of metamorphic strata is twofold, they have been deposited at one period, they have become crystalline at another. We can rarely hope to define with exactness the date of both these periods, the fossils having been destroyed by plutonic action, and the mineral characters being the same, whatever the age. Superposition itself is an ambiguous test, especially when we desire to determine the period of crystallization. Suppose, for example, we are convinced that certain metamorphic strata in the Alps, which are covered by cretaceous beds, are altered lias; this lias may have assumed its crystalline texture in the cretaceous or in some tertiary period, the

Eocene for example. If in the latter, it should be called Eocene, when regarded as a metamorphic rock, although it be liassic, when considered in reference to the era of its deposition. According to this view, the superposition of chalk does not prevent the subjacent *metamorphic* rock from being Eocene. If, however, in the progress of science, we should succeed in ascertaining the twofold chronological relations of the metamorphic formations, it might be useful to adopt a twofold terminology. We might call the strata above alluded to Liassic-Eocene, or Liassic-Cretaceous; the first term referring to the era of deposition, the second to that of crystallization. According to this method, the chlorite-schist, mica-schist, and gneiss of the Malvern Hills, would belong to the Silurian-Old Red sandstone period, because they are Silurian strata altered into metamorphic rocks during the deposition of the Old Red sandstone. (See p. 500.)

We have seen, when discussing the ages of the plutonic rocks, that examples occur of various primary, secondary, and tertiary deposits converted into metamorphic strata, near their contact with granite. There can be no doubt in these cases that strata, once composed of mud, sand, and gravel, or of clay, marl, and shelly limestone, have for the distance of several yards, and in some instances several hundred feet, been turned into gneiss, mica-schist, hornblende-schist, chlorite-schist, quartz rock, statuary marble, and the rest. (See Chapters 10. and 11.)

But when the metamorphic action has operated on a grander scale, it tends entirely to destroy all monuments of the date of its development. It may be easy to prove the identity of two different parts of the same stratum; one, where the rock has been in contact with a volcanic or plutonic mass, and has been changed into marble or horn-blende-schist, and another not far distant, where the same bed remains unaltered and fossiliferous; but when we have to compare two portions of a mountain chain — the one metamorphic, and the other unaltered — all the labour and skill of the most practised observers are required. I shall mention one or two examples of alteration on a grand scale, in order to explain to the student the kind of reasoning by which we are led to infer that dense masses of fossiliferous strata have been converted into crystalline rocks.

Northern Apennines.—Carrara.—The celebrated marble of Carrara, used in sculpture, was once regarded as a type of primitive limestone. It abounds in the mountains of Massa Carrara, or the " Apuan Alps," as they have been called, the highest peaks of which are nearly 6000 feet high. Its great antiquity was inferred from its mineral texture, from the absence of fossils, and its passage

downwards into talc-schist and garnetiferous mica-
schist; these rocks again graduating downwards
into gneiss, which is penetrated, at Forno, by granite veins.   Now the researches of MM. Savi,
Boué, Pareto, Guidoni, De la Beche, and especially Hoffmann, have demonstrated that this marble,
once suposed to be formed before the existence of
organic beings, is, in fact, an altered limestone of
the oolitic period, and the underlying crystalline
schists are secondary sandstones and shales, modified by plutonic action.   In order to establish
these conclusions it was first pointed out, that the
calcareous rocks bordering the Gulf of Spezia, and
abounding in oolitic fossils, assume a texture like
that of Carrara marble, in proportion as they are
more and more invaded by certain trappean and
plutonic rocks, such as diorite, euphotide, serpentine, and granite, occurring in the same country.

It was then observed that, in places where the
secondary formations are unaltered, the uppermost
consist of common Apennine limestone with nodules of flint, below which are shales, and at the
base of all, argillaceous and siliceous sandstones.
In the limestone, fossils are frequent, but very
rare in the underlying shale and sandstone.   Now
a gradation has been traced laterally from these
rocks into another and corresponding series, which
is completely metamorphic; for at the top of this
we find a white granular marble, wholly devoid of

fossils, and almost without stratification, in which there are no nodules of flint, but in its place siliceous matter disseminated through the mass in the form of prisms of quartz. Below this, and in place of the shales, are talc-schists, jasper, and hornstone; and at the bottom, instead of the siliceous and argillaceous sandstones, are quartzite and gneiss.* Had these secondary strata of the Apennines undergone universally as great an amount of transmutation, it would have been impossible to form a conjecture respecting their true age; and then, according to the common method of geological classification, they would have ranked as primary rocks. In that case the date of their origin would have been thrown back to an era antecedent to the deposition of the Lower Cambrian strata, although in reality they were formed in the oolitic period, and altered at some subsequent and unknown epoch.

Alps of Switzerland. — In the Alps, analogous conclusions have been drawn respecting the alteration of strata on a still more extended scale. In the eastern part of that chain, some of the primary fossiliferous strata, as well as the older secondary formations, together with the oolitic and cretaceous rocks, are distinctly recognizable. Ter-

* See Notices of Savi, Hoffmann, and others, referred to by Boué, Bull. de la Soc. Géol. de France, tom. v. p. 317 and tom. iii. p. xliv.

tiary deposits also appear in a less elevated position on the flanks of the Eastern Alps; but in the
Central or Swiss Alps, the primary fossiliferous,
and older secondary formations disappear, and
the cretaceous, oolitic, and liassic strata gra-
duate insensibly into metamorphic rocks, consisting of granular limestone, talc-schist, talcose-
gneiss, micaceous schist, and other varieties.   In
regard to the age of this vast assemblage of crystalline strata, we can merely affirm that some of
the upper portions are altered newer secondary
deposits: but we cannot avoid suspecting that the
disappearance both of the older secondary and
primary fossiliferous rocks may be owing to their
having been all converted in this region into crystalline schist.

It is difficult to convey to those who have never
visited the Alps a just idea of the various proofs
which concur to produce this conviction.   In the
first place, there are certain points where strata of
the Oolite, Lias, and Chalk have been turned into
granular marble, gneiss, and other metamorphic
schists, near their contact with granite.   This fact
shows undeniably that plutonic causes continued
to be in operation in the Alps down to a late period, even after the deposition of some of the
newer secondary formations.   Having established
this point, we are the more willing to believe that
many inferior fossiliferous rocks, probably exposed

for longer periods to a similar action, may have become metamorphic to a still greater extent.

We also discover in parts of the Swiss Alps dense masses of strata of the age of the Green-sand and Chalk, which have assumed that semi-crystalline texture which Werner called transition, and which naturally led his followers, who attached great importance to mineral characters taken alone, to class them as transition formations, or as groups older than the lowest secondary rocks. (See p. 263.) Now, it is probable that these strata have been affected, although in a less intense degree, by that same plutonic action which has entirely altered and rendered metamorphic so many of the subjacent formations; for in the Alps, this action has by no means been confined to the immediate vicinity of granite. Granite, indeed, and other plutonic rocks rarely make their appearance at the surface, notwithstanding the deep ravines which lay open to view the internal structure of these mountains. That they exist below at no great depth we cannot doubt, and we have already seen (p. 211.) that at some points, as in the Valorsine, near Mont Blanc, granite and granitic veins are observable, piercing through talcose gneiss, which passes insensibly upwards into secondary strata.

It is certainly in the Alps of Switzerland and Savoy, more than in any other district in Europe, that the geologist is prepared to meet with the signs of an intense development of plutonic action; for here we find the most stupendous monuments of mechanical violence, by which strata thousands of feet thick have been bent, folded, and overturned. (See p. 113.) It is here that marine secondary formations of a comparatively modern date, such as the oolitic and cretaceous, have been upheaved to the height of 10,000, or even 12,000 feet above the level of the sea; and even tertiary strata, apparently of the Miocene era, have been raised to an elevation of 4000 or 5000 feet, so as to rival in height the loftiest mountains in Great Britain.

If the reader will consult the works of many eminent geologists who have explored the Alps, especially those of MM. De Beaumont, Studer, Necker, and Boué, he will learn that they all share, more or less fully, in the opinions above expressed. It has, indeed, been stated by MM. Studer and Hugi, that there are complete alternations on a large scale of secondary strata, containing fossils, with gneiss and other rocks, of a perfectly metamorphic structure. I have visited some of the most remarkable localities referred to by these authors, but although agreeing with them that there are passages from the fossiliferous to the metamorphic series far from the contact of granite or other plutonic rocks, I was unable to convince myself that the distinct alternations of highly crystalline, with unaltered strata above alluded to,

might not admit of a different explanation. In
one of the sections described by M. Studer in the
highest of the Bernese Alps, namely in the Roth-
thal, a valley bordering the line of perpetual snow
on the northern side of the Jungfrau, I observed
a mass of gneiss 1000 feet thick, and 15,000 feet
long, not only resting upon, but also again covered
by strata containing oolitic fossils. These anoma-
lous appearances may partly be explained by supposing great solid wedges of intrusive gneiss to
have been forced in laterally between strata to
which I found them to be in many sections unconformable. The superposition, also, of the
gneiss to the oolite may, in some cases, be due
to a reversal of the original position of the beds in
a region where the convulsions have been on so
stupendous a scale.

On the Sattel also, at the base of the Ges-
tellihorn, above Enzen, in the valley of Urbach,
near Meyringen, some of the intercalations of
gneiss between fossiliferous strata may, I conceive,
be ascribed to mechanical derangement. Almost
any hypothesis of repeated changes of position
may be resorted to in a region of such extraordinary confusion. The secondary strata may first
have been vertical, and then certain portions may
have become metamorphic (the plutonic influence
ascending from below) while intervening strata
remained unchanged. The whole series of beds

may then again have been thrown into a
nearly horizontal position, giving rise to the superposition of crystalline upon fossiliferous formations.

It was remarked, in the last chapter, that as the
hypogene rocks, both stratified and unstratified,
crystallize originally at a certain depth beneath
the surface, they must always, before they are upraised and exposed at the surface, be of considerable antiquity, relatively to a large portion of the
fossiliferous and volcanic rocks. They may be
forming at all periods; but before any of them
can become visible, they must be raised above
the level of the sea, and some of the rocks which
previously concealed them must have been removed by denudation. If the student will refer
to the frontispiece, he will see that the strata A,
which were the last deposited, are every where
hidden from human observation by the sea, while
the contemporaneous metamorphic rocks C are
concealed at a still greater depth, as are also the
plutonic rocks D of the same age. He will also
observe that the strata C, which have recently
become metamorphic, are not parts of A, nor
even of the groups immediately antecedent in
date a, b, c, but they are portions of much older
formations, d, e, f, g, h, i. Now, suppose that part
of the earth's crust, which is represented in the
frontispiece to be subjected, in various places, to

a long series of upheaving and depressing movements; the beds A will, here and there, be partially upraised and converted into dry land, but the hypogene rocks C, D, although brought up nearer to the surface, will still, very probably, remain hidden from sight. Let a second period elapse and the rocks A may be raised in some countries to a height of several thousand feet; and still the rocks C and D may be almost every where hidden. During a third period, when the stratified formations A have been laid dry over large continental areas, and have reached the summits of some Alpine chains, the hypogene formations C D may also be forced up and exposed to view above the level of the ocean by similar causes; but they will rank no longer as modern rocks, the geologist being already acquainted with newer groups, both fossiliferous and volcanic. The student will also perceive how impossible it may then be to prove that the strata C became metamorphic at the period of the deposition of A, and how difficult not to exaggerate the antiquity of C as a series of metamorphic rocks, when the remote period of their deposition has been ascertained, and the comparatively modern era of their crystallization remains uncertain.

Order of succession in Metamorphic rocks. — There is no universal and invariable order of superposition in metamorphic rocks, although a particular arrangement may prevail throughout countries of great extent, for the same reason that it is traceable in those sedimentary formations from which crystalline strata are derived.   Thus, for example, we have seen that in the Apennines, near Carrara, the descending series, where it is metamorphic, consists of, 1st, saccharine marble; 2dly, talcose-schist; and 3dly, of quartz-rock and gneiss; where unaltered, of, 1st, fossiliferous limestone; 2dly, shale; and 3dly, sandstone.

But if we investigate different mountain chains we find gneiss, mica-schist, hornblende-schist, chlorite-schist, hypogene limestone, and other rocks, succeeding each other, and alternating with each other, in every possible order.   It is, indeed, more common to meet with some variety of clay-slate forming the uppermost member of a metamorphic series than any other rock; but this fact by no means implies, as some have imagined, that all clay-slates were formed at the close of an imaginary period, when the deposition of the crystalline strata gave way to that of ordinary sedimentary deposits.   Such clay-slates, in fact, are variable in composition, and sometimes alternate with fossiliferous strata, so that they may be said to belong almost equally to the sedimentary and metamorphic order of rocks.   It is probable that had they been subjected to more intense plutonic action, they would have been transformed into hornblende-schist, foliated chlorite-schist, scaly talcose-schist, mica-schist, or other more perfectly crystalline rocks, such as are usually associated with gneiss.

Uniformity of mineral character in Hypogene rocks.—Humboldt has emphatically remarked that when we pass to another hemisphere, we see new forms of animals and plants, and even new constellations in the heavens; but in the rocks we still recognize our old acquaintances, — the same granite, the same gneiss, the same micaceous schist, quartz-rock, and the rest.   It is certainly true that there is a great and striking general resemblance in the principal kinds of hypogene rocks, although of very different ages and countries; but it has been shown that each of these are, in fact, geological families of rocks, and not definite mineral compounds.   They are much more uniform in aspect than sedimentary strata, because these last are often composed of fragments varying greatly in form, size, and colour, and contain fossils of different shapes and mineral composition, and acquire a variety of tints from the mixture of various kinds of sediment.   The materials of such strata, if melted and made to crystallize, would be subject to chemical laws, simple and uniform in their action, the same in every climate, and wholly undisturbed by mechanical and organic causes.

Nevertheless, it would be a great error to assume that the hypogene rocks, considered as aggregates of simple minerals, are really more homogeneous in their composition than the several members of the sedimentary series. In the first place, different assemblages of hypogene rocks occur in different countries; and secondly, in any one district, the rocks which pass under the same name are often extremely variable in their component ingredients, or at least in the proportions in which each of these are present. Thus, for example, gneiss and mica-schist, so abundant in the Grampians, are wanting in Cumberland, Wales, and Cornwall; in parts of the Swiss and Italian Alps, the gneiss and granite are talcose, and not micaceous, as in Scotland; horn-blende prevails in the granite of Scotland — schorl in that of Cornwall — albite in the plutonic rocks of the Andes — common felspar in those of Europe. In one part of Scotland, the mica-schist is full of garnets; in another it is wholly devoid of them: while in South America, according to Mr. Darwin, it is the gneiss, and not the mica-schist, which is most commonly garnetiferous. And not only do the proportional quantities of felspar, quartz, mica, hornblende, and other minerals, vary in hypogene rocks bearing the same name; but what is still more important, the ingredients, as we have seen, of the same simple mineral are not always constant.   (p. 147., and table, p. 166.)

The Metamorphic strata, why less calcareous than the fossiliferous. — It has been remarked, that the quantity of calcareous matter in metamorphic strata, or, indeed, in the hypogene formations generally, is far less than in fossiliferous deposits. Thus the crystalline schists of the Grampians in Scotland, consisting of gneiss, mica-schist, hornblende-schist, and other rocks, many thousands of yards in thickness, contain an exceedingly small proportion of interstratified calcareous beds, although these have been the objects of careful search for economical purposes.   Yet limestone is not wanting in the Grampians, and it is associated sometimes with gneiss, sometimes with mica-schist, and in other places with other members of the metamorphic series.   But where limestone occurs abundantly, as at Carrara, and in parts of the Alps, in connection with hypogene rocks, it usually forms one of the superior members of the crystalline group.

The scarcity, then, of carbonate of lime in the plutonic and metamorphic rocks generally, seems to be the result of some general cause. So long as the hypogene rocks were believed to have originated antecedently to the creation of organic beings, it was easy to impute the absence of lime to the non-existence of those mollusca and zoophytes by which shells and corals are secreted; but when we ascribe the crystalline formations to plutonic action, it is natural to inquire whether this action itself may not tend to expel carbonic acid and lime from the materials which it reduces to fusion or semi-fusion. Although we cannot descend into the subterranean regions where volcanic heat is developed, we can observe in regions of spent volcanos, such as Auvergne and Tuscany, hundreds of springs both cold and thermal, flowing out from granite and other rocks, and having their waters plentifully charged with carbonate of lime. The quantity of calcareous matter which these springs transfer, in the course of ages, from the lower parts of the earth's crust to the superior or newly formed parts of the same, must be considerable.*

If the quantity of siliceous and aluminous ingredients brought up by such springs were great, instead of being utterly insignificant, it might be contended that the mineral matter thus expelled implies simply the decomposition of ordinary subterranean rocks; but the prodigious excess of carbonate of lime over every other element must, in the course of time, cause the crust of the earth below to be almost entirely deprived of its calcareous constituents, while we know that the same

* See Principles, *Index*, " Calcareous Springs."

action imparts to newer deposits, ever forming in seas and lakes, an excess of carbonate of lime. Calcareous matter is poured into these lakes and the ocean by a thousand springs and rivers; so that part of almost every new calcareous rock chemically precipitated, and of many reefs of shelly and coralline stone, must be derived from mineral matter subtracted by plutonic agency, and driven up by gas and steam from fused and heated rocks in the bowels of the earth.

Not only carbonate of lime, but also free carbonic acid gas is given off plentifully from the soil and crevices of rocks in regions of active and spent volcanos, as near Naples, and in Auvergne. By this process, fossil shells or corals may often lose their carbonic acid, and the residual lime may enter into the composition of augite, hornblende, garnet, and other hypogene minerals. That the removal of the calcareous matter of fossil shells is of frequent occurrence, is proved by the fact of such organic remains being often replaced by silex or other minerals, and sometimes by the space once occupied by the fossil being left empty, or only marked by a faint impression. We ought not indeed to marvel at the general absence of organic remains from the crystalline strata, when we bear in mind how often fossils are obliterated, wholly or in part, even in tertiary formations—how often vast masses of sandstone and shale, of different ages,

and thousands of feet thick, are devoid of fossils —
how certain strata may first have been deprived of
a portion of their fossils when they became semi-
crystalline, or assumed the *transition* state of Werner — and how the remaining organic remains
may have been effaced when they were rendered
metamorphic.   Some rocks of the last-mentioned
class, moreover, must have been exposed again
and again to renewed plutonic action.

INDEX.

A.

ABERDEENSHIRE, granite of, 203.
Acephalous mollusca, 60.
Acrodus nobilis, 389.
Actinoceras Simmsii, 463.
Actinolite, 166. 224.
Agassiz, on fossil fish, 309. 388, 389. 409, 416. 426. 454.
Age of aqueous strata, how determined, 271.
——, of volcanic rocks, 467.
——, of the plutonic rocks, 487.
——, of the metamorphic rocks, 509.
Airdnamurchan, trap veins in, 171.
Albite, 166.
Alluvium described, 131.
——, passes into regular strata, 133.
——, marine, 135.
Alps, reversed position of strata in, 113. 518.
——, curved strata of, 114.
——, metamorphic rocks of the, 497. 515.
Altered rocks, 17. 175. 204. 235. 241. 495. 509.
Alternations of coarse and fine strata, how formed, 8. 33.
——, of marine and freshwater formations, 68.
Alumine in rocks, how to detect, 28.
Amblyrhynchus cristatus, 395.
America, Recent and Tertiary strata of, 295.
—— Silurian strata in, 462.
Amici, Professor, on recent Charæ, 67.
Ammonites, figures of, 327. 379. 410. 425.
Ampelite, 225.
Amphibolite, 161. 225.
Ampullaria glauca, 64.
Amygdaloid described, 155.

Ananchytes ovatus, 318.
Ancylus elegans, 62.
Andes, geological structure of, 494.
——, tertiary plutonic rocks of, 495.
Anglesea, rocks altered by a dike in, 175.
Anodonta, figures of, 61.
Anoplotherium, 311.
Ansted, Mr., on Cambrian fossils, 465.
Anticlinal line explained, 101. 110.
Antrim, rocks altered by dikes in, 175.
——, on age of trap rocks of, 480.
Apennines, age of metamorphic rocks of, 513.
Apes, fossil, 311.
Apiocrinites rotundus, 373.
Aqueous rocks described, 5. 271.
Arbroath, section from, to the Grampians, 99.
Arenaceous rocks described, 26.
Argillaceous rocks described, 27. 223.
Arran, dikes in, 171.
——, geology of, 506.
——, section of, 507.
Arthur's seat, strata altered in, 179,
Asaphus, figures of, 459. 461.
Ashby, faults in coal-field of, 128.
Ashes, volcanic, hollows filled up by, 40.
——, wide dispersion of, 472.
Astarte, 305.
Atlantis, 363.
Auch, ape fossil near, 312.
Augite and hornblende, analogy of, 148.
——, analysis of, 166.
Augite rock, 161.
Augitic porphyry, 161.
Auricula, 62.
Autreppe, unconformable strata, 115.
Auvergne, volcanos of, 11. 145. 477.
——, fresh water strata of, 58.

A A

THE END.

LONDON:
Printed by A. SPOTTISWOODE,
New-Street-Square.